Intertemporal Resource Economics

Karl Farmer • Birgit Bednar-Friedl

Intertemporal Resource Economics

An Introduction to the Overlapping Generations Approach

Springer

Prof. Dr. Karl Farmer
Dr. Birgit Bednar-Friedl
University of Graz
Economics
Universitätsstrasse 15/F4
A-8010 Graz
Austria
karl.farmer@uni-graz.at
birgit.friedl@uni-graz.at

ISBN 978-3-642-13228-5 e-ISBN 978-3-642-13229-2
DOI 10.1007/978-3-642-13229-2
Springer Heidelberg Dordrecht London New York

Library of Congress Control Number: 2010931865

Cover design: WMX Design

Printed on acid-free paper

Springer is part of Springer Science+Business Media (www.springer.com)

Preface

This book represents both a textbook and a monograph. Part I entitled 'Basics' mainly contains textbook material and this is also partly true for Part II. Here the precise definition and characterization of the not so well-known notion of intergenerational efficiency within Diamond's two-period overlapping generations model figures prominently. In Part III, a renewable natural resource is introduced in the log-linear Cobb-Douglas overlapping generations model, and the efficiency concepts developed in Part II are applied. The balance among material for a textbook und for a monograph is approximately even. In Part IV research monograph characteristics gain progressively prominence. While this part dealing with intergenerational equity in perfectly competitive market economies presents already published work, the last part focusing on harvest cost contains still unpublished work.

As the subtitle of the present book announces, our intention is to provide an introduction to the not so widespread overlapping generations approach to intertemporal resource economics. It is introductory in that utility and production functions are functionally specified such that the interested reader can derive explicit solutions to intertemporal general equilibria. However, we do not primarily aim at enhancing the reader's skill of solving general equilibrium models—we rather aim at providing the tools for coping with analytically much more advanced dynamic general equilibrium models with renewable natural resources, published in leading journals.

This book emerged out of lectures of the first author within the master programs of the University of Life Sciences in Vienna and at Karl-Franzens-University of Graz. Parts IV builds on already published work of the first author while the last Part contains joint work of both authors. The responsibility for the content of Parts I - IV lies primarily with the first author, while both authors share responsibility for Part V.

Several people have helped us during the gestation of this book. First, we are grateful to Reinhard Wurzinger who typeset the lecture notes in LaTeX. Second, special thanks go to Laurie Conway for his excellent language check. We are also grateful to comments of students in Vienna and Graz, especially to Rosa Lugger which helped us to eliminate errors and passages which were difficult to understand. Last but not least we thank Dr. Werner Müller and Barbara Fess from Springer for accepting the manuscript for publication and for providing encouragement and support during the writing process.

The biggest debt, however, owes the first author to his family and the second to her husband for patience and support during the months necessary for completing the manuscript.

Karl Farmer
Birgit Bednar-Friedl
University of Graz, April 2010

Contents

Part I Basics

Part I
Basics

Chapter 1
Introduction

1.1 Motivation

Conventional belief holds that the ever globalizing market economy is running out of natural resources, and hence is doomed to fail in the foreseeable future (see Meadows et al, 1972, 1992, for a system–dynamic foundation of the conventional belief). Long–run resource statistics however show exactly the opposite trend: natural resource scarcity as measured by real unit extraction costs of almost all fossil and mineral resources is decreasing (see the classic study by Barnett and Morse, 1963). Although more recent empirical studies of real resource prices (see Slade, 1982; Hamilton, 2008; Ghoshray and Johnson, 2010) have also detected rising prices for some natural resources, the empirical evidence on increasing resource scarcity remains mixed (Endres and Querner, 2000, 18). The question thus arises whether natural resource scarcity is an imaginary or a real problem in economics. We will address this major question in the present book by dividing it into subquestions and respective brief answers to provide an intellectual roadmap to the reader of this book.

First, which sort of scarcity is typical for natural resources? Indeed, natural resource scarcity results from competition of subsequent generations for the limited amount of natural resources. Second, does this intergenerational scarcity depend on the resource examined as well as on the time period considered? This is in fact the case and, to address the competition for resources between generations, we will focus on natural resources that outlive the agents who use it.

Third, how can conventional economic knowledge, that scarcity is alleviated by an efficient use of resources, be applied to the solution of the intergenerational allocation problem? We will show that the well-known static Pareto-efficiency notion can be adapted to the intergenerational dimension of the allocation problem and that it moreover applies to all sort of scarce resources, one of which are natural resources.

Fourth, does a competitive market system provide an efficient consumption allocation for all generations? The answer to this question will mainly depend on whether we consider short-run or long-run intergenerational efficiency. Fifth, if we

K. Farmer and B. Bednar-Friedl, *Intertemporal Resource Economics: An Introduction to the Overlapping Generations Approach*, DOI 10.1007/978-3-642-13229-2_1,

focus on short-run intergenerational efficiency, how can efficient paths of harvesting renewable resources be characterized, and will self-interested agents guided by prices formed on perfectly competitive markets harvest intergenerationally efficiently? It will be shown that the short-run intergenerational efficiency of competitive market allocations depends mainly on the property rights regime with respect to the natural resource stock.

Sixth, can intergenerational equity, most often referred to as sustainability, be utilized as alternative benchmark for competitive market systems? Here the answer depends on the regeneration technology, namely whether natural resources are linearly or non-linearly regenerating. Seventh, how do shocks to the resource technology impact on economic dynamics? It turns out that the adjustment triggered by a shock to resource harvesting or to resource generation depends on the type of resource technology shock.

The rest of this chapter is devoted to an overview on the role of natural resources in market economies, as well as on the development of the field of resource economics since classical economic writers. Furthermore, we will characterize briefly the type of model we will apply throughout the book before giving an overview over remaining parts.

1.2 Natural Resources and the Economic Production Process

Natural resources are defined as those factors of production principally provided by nature. Factors of production represent the inputs into the production process, the output of which comprises commodities used to satisfy direct wants (final goods) and indirect wants (intermediate goods). In commodity production new mass is not generated but only transformed by the way of changes in material concentration and composition in space and time.

A typical distinction is made between renewable resources, i.e. resources which grow over time, and non–renewable resources, which do not. In the older literature, non–renewable resources are called 'exhaustible', but since overutilization can also exhaust a renewable resource, the more accurate term 'non–renewable' is now used. Another common distinction made is that between stock resources (such as minerals and fossil fuels) and flow resources (such as solar energy, crops). According to Ströbele (1987), it is also useful to differentiate in terms of whether the resource can be recycled (minerals) or not (fossil fuels). The above considerations thus lead us to five types of natural resources:

- solar energy, as an example of a flow resource,
- fossil energy (oil, gas, coal, uranium), as an example of a non-renewable stock resource which cannot be recycled,
- minerals (copper, iron, aluminium), as an example of a non-renewable stock resource which can be recycled,

- natural production and waste processes within ecological circular flows utilized in agriculture, forest and fishery industries (grain, timber, fish,...), as examples of a renewable (i.e. regenerating) stock resource, and
- land.

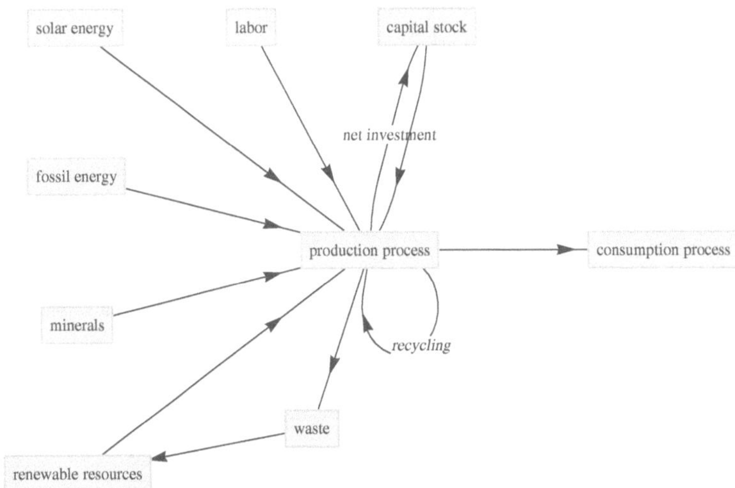

Fig. 1.1 Natural resources within the economic production process (adapted from Ströbele, 1987, 3)

This book focuses mainly on natural resources which are generated by the production and waste utilization of nature. They are called renewable or regenerating natural resources. Timber, grain, fish and environmental media are typical examples.

The natural production processes in which ordered structures of life are generated by the use of minerals, water and solar energy are self-sustaining since the production processes tend towards stable biological equilibria. While nature does not need human support, mankind depends on the fruits of natural production processes to survive.

This entails both a scientific and an economic problem. The questions that need to be posed are:

- To what extent can renewable resources be utilized by mankind without their regeneration capacity being damaged?
- Within the 'Great Society' (Hayek, 1973) of 6 to 7 billion, how intensively may natural resources be used both those with limited (i.e. exclusively defined property rights) as well as those with open access (i.e. without defined property rights)?

Before we are able to answer these questions, it is useful to classify natural resources according to four criteria: (i) how they are produced by nature, (ii) whether they can be recycled, (iii) how large are the costs of the implementation and control of exclusive property rights, and (iv) where they are used within the economy. Applying these criteria, Table 1.1 provides a typology of natural resources.

Table 1.1 Classification of natural resources

Natural production mode	Renewable (timber, fish)	Non–renewable (energy resources like oil, coal)
Recyclability	Usable once (agricultural products, energy resources)	Reusable several times (minerals)
Costs of implementation and control of exclusive individual property rights	Prohibitive costs (open–access resources)	Negligible costs (privately owned land)
Use within the economy	Resource as consumption good (oxygen in the air, fish)	resource as production input (coal, oil)

Adapted from Ströbele (1987, 5–12).

1.3 Natural Resources in the History of Economic Thought

Intertemporal resource economics dates back to the economic thought of classical economists Adam Smith, Thomas Robert Malthus and David Ricardo. The main natural resource that classical economists and their precursors, the physiocrats, focus on is agricultural land. This was a logical choice since in the pre-industrial era most people earned their subsistence from agriculture.

1.3.1 Classical Economics

For the physiocrats as precursors of classical economists, agriculture alone is capable of generating an economic surplus. It is mother nature, i.e. soil fertility, which generates the surplus. Hence, agricultural land is the natural resource considered by the physiocrats.

With the advent of manufactures and early industrial production agricultural land was no longer the sole origin of economic surplus. Industrial labor became the new source of economic surplus.

Similar to the physiocrats, classical economists considered agricultural land as a natural resource but to a lesser extent. 'Classical political economy stressed the power of the market to stimulate both growth and innovation, but remained essentially pessimistic about long-run growth prospects.' (Pearce and Turner, 1990, 6).

They thought that population growth and diminishing returns to agricultural land would reduce the profit rate and the rate of capital accumulation until a stationary state with zero rate of capital accumulation is reached.

In 1776, Smith (1723 - 1790) published *An inquiry into the nature and the causes of the wealth of nations* in which he developed a theory of 'natural' ('normal') prices determined by the prices of labor, capital and land. Land used for agricultural purposes commands a scarcity rent which accrues to the non–productive landlords.

Denoting the net product by Y, labor input by L, capital input by K, land input by R and the production function by $F(.)$, Smith's aggregate production function reads as follows:

$$Y = F(L, K, R). \tag{1.1}$$

Denoting the growth rate of the net product by g^Y, the growth rate of labor by g^L, the rate of capital accumulation by g^K, time by t, the Smithian growth equation can be written as follows:

$$g^Y = f(g^L, g^K, t).$$

It is important to notice that the growth function $f(\cdot)$ is time dependent. In contrast to neoclassical growth theory (see Solow, 1956), Smith's growth function follows no exogenously fixed pattern but emerges endogenously out of the division of labor in the capitalist system of commodity production. Under given growth rate of labor and capital, the division of labor increases output growth as long as sufficient productive land is available. Growth of the net product comes to a halt when land becomes so scarce that the diminishing returns of agricultural land outweigh the increasing returns from the division of labor. Increasing returns to scale turn into decreasing returns to scale. Average costs no longer decrease with larger output but increase. With output prices calculated on the basis of constant mark-ups on average costs, profit rates fall approaching the stationary state of zero profit and capital accumulation ceases.

In his *An essay on the principles of population* published in 1798, Malthus (1766 - 1834) claimed that food production follows the rule of an arithmetic series while population develops according to the law of a geometric series. Denoting food production in period t by Y_t, the evolution of food production over time can then be written as follows:

$$Y_t = (1/2)t\{2Y_0 + (t-1)\Delta Y\}, \tag{1.2}$$

where Y_0 denotes the initial output, and ΔY is the constant yearly increase in output. Since both Y_0 and $(t-1)\Delta Y$ (i.e. the final term in the sum of an arithmetic series) are exogenously given, it is apparent from Fig. 1.2 that Y_t is a linear function of time.

In contrast, population growth is of the exponential type, or in formal terms:

$$L_t = L_0\left(1 + g^L\right)^t \equiv L(0)e^{g^{L}t}. \tag{1.3}$$

Equation 1.3 is depicted in Fig. 1.3 where it is easy to see that population growth outperforms food production. As a consequence, people starve and population growth becomes negative until a stationary level of population is reached which can be fed by subsistence agriculture.

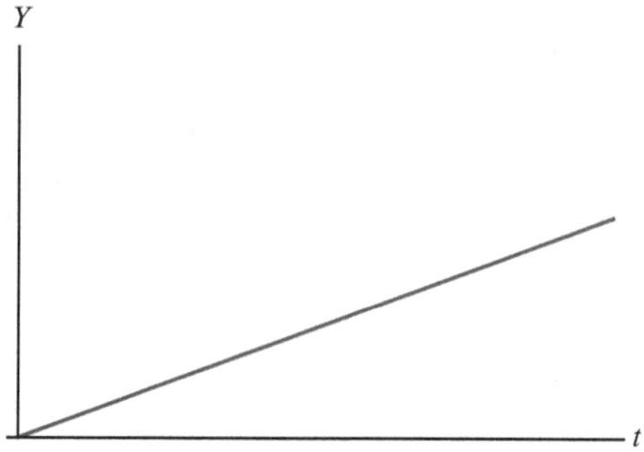

Fig. 1.2 Arithmetic growth of food production

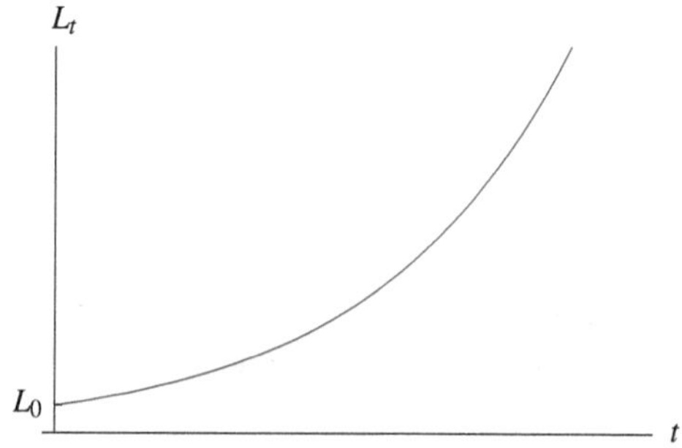

Fig. 1.3 Exponential growth of population

Malthus has thus to be credited with the insight of exponential population growth. Exponential growth of food production however escaped his notice since he—as did other classical economists—overlooked the impact of the steam engine and the associated exploitation of fossil fuels such as coal, which allowed for a subsequent exponential growth in food production.

Building on Smith's theory of natural prices, Ricardo (1772 - 1832) introduced in his major work *On the principles of political economy and taxation* (1817) the notion of *differential* rent on land of higher quality compared to a zero rent on marginal land. Hence, in contrast to Smith's aggregate production function with a constant quality of land, in Ricardo's aggregate production technology the quality of land $q(R)$ varies inversely with the amount of land already in use:

$$Y = F\left(L, K, q(R)R\right), \quad \frac{\partial q}{\partial R} < 0. \tag{1.4}$$

In the course of economic growth, successively less productive land is cultivated until marginal land is reached. Since marginal land does not command any rent, the natural price of a commodity produced by land does not include a rent component in marked contrast to Smith's theory of natural price where it does. Differential rent is thus not part of long-run production costs and hence of natural price.

Ricardo's rent and growth theory are illustrated in Fig. 1.4 which shows output as a function of population. Here marginal productivity of labor (dY/dL) determines profits and rent. For a population of L_1, workers are paid minimum wages w_{min}, land owners receive rent, and capitalists receive profits. As land of lower and lower productivity is used the higher the population, the larger become rents on high quality land and the lower become profits. Since profits are the only source of capital accumulation, output growth, and hence population growth, eventually comes to a halt.

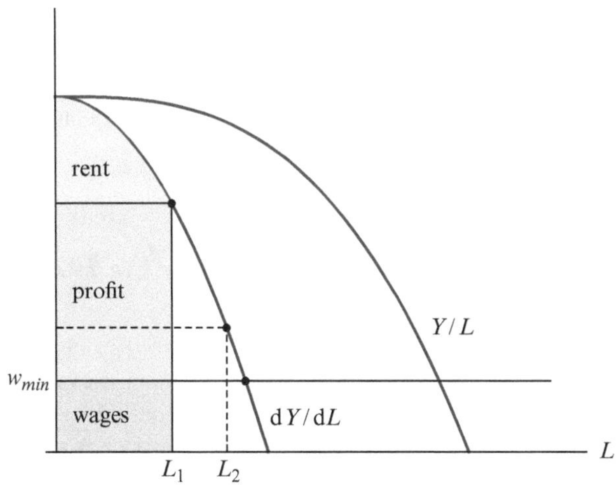

Fig. 1.4 Ricardian rent theory

This insight that (marginal) land does not receive a payment profoundly influenced neoclassical writers. Thus, in neoclassical economics natural resources no longer appear as an input in the aggregate production function because marginal land as representative of natural resources does not have any price.

1.3.2 Neoclassical Economics

As is well known, the neoclassical epoch commences with the writings of three famous economists from Switzerland, Great Britain and Austria. These were L. Walras (1834 - 1910) with his major work *Elements of pure economics* (1874), W.S. Jevons (1835 - 1882) with his *Theory of political economy* (1871), and C. Menger (1840 - 1921) with *Grundsätze der Volkswirtschaftslehre* (1871). As a result of Ricardo's theoretical lead (explicitly acknowledged by Walras in the introduction to the *Elements of pure economics*) and the impacts of early (nineteen century's) economic growth, natural resources were almost completely neglected by neoclassical economists until the end of the 1960s.

However there is no rule without exception. The German economist Faustmann developed optimal reforestation rules in 1849. The American economist Gray (in 1914) can be credited as the first to develop a systematic approach to the economics of exhaustible resources. The modern approach can be traced back to H. Hotelling's path-breaking work on the *Economics of exhaustible resources* (1931).

As in China and India today, in the 1950s and 1960s, western and middle continental Europe saw vigorous economic growth and resource conservation disappeared from the individual and political agenda.

By the end of the high-growth era in the late 1960s environmental degradation and resource exhaustion had became more apparent and the first Club of Rome report, compiled by D.H. Meadows et al (1972) and entitled *The Limits to growth*, aroused enormous political interest and also led to the foundation of modern resource (and environmental) economics.

1.3.3 Resource Economics and Politics since the 1970s

Economists are often accused of not dealing with real word problems and of concentrating too heavily on theoretical modeling of no practical relevance. Nevertheless, following the first oil price shock in 1973, several prominent economists investigated the question of whether never ending economic growth is at all feasible given finite exhaustible resources. The results of the subsequent symposium on 'economic growth and exhaustible resources' were published in a special issue of the Review of Economic Studies in 1974 and were seen at the time as the answer of the economic profession to the doomsday or neo–Malthusian forecasts of Meadows et al (1972).

With the fall of oil prices in the 1980s (after their hike in 1979 due to the Iran crisis), some economists such as J. L. Simon and H. Kahn (1984) appeared on the scene and began to stress the almost unlimited potential of natural resources. Counter arguments also appeared, for example from R. Repetto (1985) at the World Watch Institute who pointed to clear limits on both resources and the environment. A fundamental change in perspective also took place in that economic growth was perceived to be a great risk for renewable rather than exhaustible resources. While the latter tend to be protected from overexploitation by the presence of property rights, such rights are often absent from renewables. The resulting lack of market incentives thus leads to relatively greater vulnerability.

If intertemporal and intergenerational efficiency cannot be adequately ensured by the market system, it is only natural to ask which benchmark should be used in practice in order to govern private resource utilization. The answer was provided in the so-called Brundtland report entitled *Our common future* (WCED, 1987, 40). 'Sustainable Development' is here defined as development which 'seeks to meet the needs and aspirations of the present without compromising the ability to meet those of the future.' Not surprisingly, sustainable development so defined, focuses on intergenerational equity rather than on intergenerational efficiency. Moreover, from an ecological perspective, here sustainable development also implies the conservation of essential natural resources, which is clearly at odds with current utilization trends of exhaustible and renewable resources.

Against these notions of 'strong sustainability' authors like Simon (1996) in the *Ultimate resource 2* and Lomborg (2001) in *The sceptical environmentalist* argue that human ingenuity has always enabled man to overcome natural resource scarcity and there is no reason why this should not be the case in the future. While we accept that such authors are right to combat empirically unfounded pessimism of entrenched environmentalists and resource conservationists, in our analysis we prefer to leave the long-run perspective open and by focusing on intertemporal general equilibrium approaches, we merely aim to identify those sufficient conditions which imply a 'good' or a 'bad' long-run outcome. Thus, the question we address is not whether economic development ultimately leads to resource collapse or whether economic development can deal with relative resource scarcity. We focus instead on the specific assumptions and conditions which result in the different outcomes.

1.4 Resource Scarcity, Market Equilibrium, Intergenerational Efficiency and Equity

The basic problem of resource economics is establishing the best use of limited natural resources by subsequent human generations. From this perspective resource economics focuses thus on an *intergenerational* allocation problem. This intergenerational dimension of the problem was also stressed by the well-known natural resource and environmental economists Pearce and Turner, 'For all we know, it may be perfectly possible to dispense with natural environments in favor of an encapsulated world of plastics and microchips. But the issue surely is one of how far down that road we wish to travel, and what regrets we think our grandchildren will have if we travel down it too far.' (Pearce and Turner, 1990, xii)

In this section, we review the fundamental concepts we will use in the subsequent analysis, namely resource scarcity, intergenerational efficiency and equity and how they can be achieved in competitive market systems.

1.4.1 Intergenerational Scarcity of Renewable Resources

There is a widely held view that non-renewable resources are scarce, i.e. that the demand for non-renewable resources is larger than the supply, while renewable resources are not, since they regenerate themselves. However, this view involves a misunderstanding in that scarcity is defined with respect to an absolute amount. In reality, scarcity comes about when demand is large relative to supply. For example, if at some point in time the demand for a renewable resource is larger than the supply coming from natural regeneration and harvesting, the resource is scarce, and society has to decide which of the competing resource uses are to be satisfied and which not.

If contemporary and subsequent generations can be said to be in competition for the use of renewable resources, resource scarcity means that the needs of every generation in every period can not be satisfied in spite of natural regeneration. Hence, the real question is how the present generation may harvest renewable resources such that sufficient regeneration capacity remains intact for future generations. Since not all generations can be satisfied simultaneously it is rational to figure out an allocation which satisfies the needs of subsequent generations as much as possible. Following Pareto (1909), a mathematical solution to this problem of intergenerationally efficient allocation has been found. This is employed, together with an abstract market solution, in the next subsection.

1.4.2 Intergenerational Efficiency and Intertemporal Market Equilibrium

Before delving more deeply into the intertemporal allocation problem, it is useful to distinguish natural resources according to their respective time horizon of regeneration. A truly intertemporal allocation problem arises only for resources which regenerate from one year to the next. In Table 1.2 natural resources are differentiated in terms of time span of natural regeneration.

Table 1.2 Natural resources and their horizon of regeneration

Time horizon of regeneration	Period of utilization	Examples	Theoretical concept
<1 year	Immediately after ripening	Fruits, corn	Static market model
1–150 years	Following intertemporal optimization	Fish, forest	Theory of renewable resources
100–1000 years	Following intertemporal optimization	Atmosphere (CO_2), biodiversity	Theory of renewable resources
>1000 years	Following intertemporal optimization	Copper, crude oil, coal, ...	Theory of exhaustible resources

Adapted from Ströbele (1987, 13).

Roughly speaking, a resource harvesting path is *intergenerationally efficient*, if there is no other harvesting path along which one generation achieves a higher utility level and no other generation suffers a utility loss. It can be shown (see, e.g. Varian, 1986) that this definition is equivalent to the mathematical problem of maximizing the utility of the present old generation subject to the constraints that the present and subsequent future generations obtain certain utility levels. It is important to acknowledge from the outset that in current parliamentary democracies no institutions exist to ensure intergenerationally efficient allocation. Intergenerational efficiency is a benchmark concept against which real mechanisms for allocating renewable resources across different generations may be compared, e.g. market or electoral mechanisms.

Markets in societies with an extended division of labor and an associated dispersion of knowledge among individuals (Hayek, 1945) (1) induce individuals to transmit their subjective knowledge to other individuals in need of this knowledge and (2) control subjective errors in knowledge acquisition and transmission among individuals (Hayek, 1979). While in reality markets and entrepreneurial competition are clearly indispensable in directing knowledge acquisition through time, mainstream economics normally tends to assume that the knowledge problem has already been solved and simply aims to provide a mathematical model of market economies focusing on balances of supply and demand across all markets (general market equilibrium).

As part of mainstream economics, intertemporal resource economics thus works with the mathematical model of an *intertemporal market equilibrium* in which households and firms maximize their utility and profit functions over more than one period and in which markets for factors and products clear in every period. Typical questions here concern an analysis of the existence and stability of intertemporal equilibria, including the possibility of steady state solutions. Once assured of the existence and stability of intertemporal market equilibria the main question becomes: Under which conditions do intertemporally optimal individual harvesting decisions, coordinated by price systems extending over time, lead to an intergenerationally efficient allocation. In particular, is a steady-state market equilibrium long-run intergenerationally efficient?

1.4.3 Intergenerational Equity (Sustainability) versus Intergenerational Efficiency

At the latest, since the publication of the Brundtland report, the idea of intergenerational efficiency as the main benchmark in evaluating real world allocation mechanisms has been severely questioned. The problem is that intergenerational efficiency is compatible with an extremely unequal distribution of utilities across subsequent generations. Even though classical economists like J. S. Mill and K. Marx had dealt with the concept of intergenerational equity, it did not figure prominently in economic reasoning until the Brundtland report popularized the idea in the form of 'sustainability' and suggested it to be used as substitute for or together with intergenerational efficiency.

As a result of political differences between advanced and developing countries the sustainability notion expressed in the Brundtland report had to remain relatively vague. It was left to (economic) science to clarify the concept. Following Hanley et al (2007), two different notions of sustainable development (sustainability) can be distinguished: the 'outcome approach' and the 'opportunities approach'. According to the outcome approach, development is *sustainable* if the utility of subsequent generations does not decline over time. The second approach focuses on the means (capital) which are available to society to generate utility (welfare). On distinguishing four types of capital, namely man-made capital, human capital, natural capital (i.e. natural resources), and social capital, development is said to be sustainable if one or more types of capital does not decline over time.

Irrespective of which sustainability notion is used it is obvious that intergenerational efficiency and sustainability are different concepts. Less obvious is the relationship between the two concepts. For most economists (see e.g. Farzin, 2006) according to the outcome approach intergenerational efficiency is a necessary condition for sustainability. Another question here is whether the 'deep' parameters of decentralized market economies allow for the existence of sustainable development in the sense that natural capital does not decline over time (see Mourmouras, 1991; Farmer, 2000). To answer these and other questions analytically, we need fully specified intertemporal equilibrium models. The sort of equilibrium models we use is described in the next section.

1.5 General Equilibrium Models

Following mainstream resource economics (see e.g. Dasgupta and Heal, 1979; Conrad and Clark, 1987; Pearce and Turner, 1990; Neher, 1990; Conrad, 1999; Perman et al, 2003), we tackle the problem of optimal resource utilization by utilizing mathematical models representing equilibria between supply of and demand for economic goods and natural resources.

In equilibrium models the notion of partial equilibrium is to be distinguished from that of general equilibrium. In resource economics, partial equilibrium models, e.g. addressing optimal harvesting of a fish stock or a forest under different property rights regimes, predominate. Despite the great detail which can be included in this partial equilibrium approaches, the question of resource scarcity and the resulting implications for sustainable development cannot be properly addressed within such a framework. Considering that natural resources (like oil) are, similar to labor, an input into the production of not only a small set of commodities but of the whole set of commodities produced in modern economies, resource scarcity can pose a fundamental threat to economic growth in the long run. Thus, we focus here on *general equilibrium models* where natural resources are analyzed in their role for commodity production and consumption.

Among general equilibrium models, *temporary* (i.e. one-period) and *intertemporal* general equilibrium models are to be distinguished. As mentioned above, in this book we focus on natural resources which are more than one period in existence. As a consequence, the agents in our general equilibrium models are confronted with intertemporal decision problems. The intertemporal nature of the main decision problem is described in the next subsection.

1.5.1 Intertemporal General Equilibrium Models

Intertemporal (general) equilibrium models deal with the optimal use of natural resources and economic goods *over time* whereby current use constrains future utilization. As usual, optimality begs the question of for whom the use of a natural resource is optimal. In our basic model both the resource owner (if any) and a central planner optimize. The former optimizes in order to satisfy his or her own interests, while the latter is assumed (first by Pigou, 1920,1962) to act as an agent of as yet unborn future generations.

Intertemporal optimality (efficiency) resolves the trade-off between the current and future use of a natural resource in limited supply in both the present and the future. Or in other words: How much should the owner of the resource or the central planner use up now at the expense of use tomorrow?

There are two main types of intertemporal general equilibrium models: the model with overlapping generations (OLG) and the model with infinitely lived agents (ILA). Their commonalities and differences are dealt with in the next subsection.

1.5.2 Overlapping Generations versus Infinitely Lived Agents

As the name suggests, in the intertemporal general equilibrium model with one or many *infinitely lived agents* the planning horizon of the agent(s) equals the number of time periods the economy exists: namely infinity. There is no generation overlap. In most ILA models there is only one agent who is representative for all agents in the economy.[1]

In contrast, in intertemporal equilibrium models with *overlapping generations* the planning horizon of agents comprises a smaller number of periods than the economy exists. In the now classical version (Samuelson, 1958; Diamond, 1965), agents live for two periods and overlap for one period. As a consequence, the economy and the natural environment are long-lived, while the agents are short-lived. Natural resources serve, either as substitute for or in addition to man-made capital, as important store of value between generations. This basic demographic setting thus provides us with a structure in which intergenerational conflict and the limited concern of present generations for future generations can generate problems of resource overuse and a decline of consumption over time.

[1] For an overview on ILA models with renewable resources, see e.g. Clark (1990); Johannson and Mäler (1985).

1.5.3 The Intergenerational Conflict and the Lack of Property Rights

The OLG structure allows for outcomes in the economic process in line with the 'litany' described by Lomborg (2001, 4):

'Our resources are running out. The population is ever growing, leaving less and less to eat. [...] The planet's species are becoming extinct in vast numbers - we kill off more than 40,000 each year. The forests are disappearing, fish stocks are collapsing and the coral reefs are dying.'

While Lomborg (2001, 4) tries to show by careful statistical analysis that this standard litany 'does not seem to be backed up by the available evidence', many environmentalists and other 'friends of the earth' remain unconvinced. To their mind, the capitalist system is governed by a morally unchecked profit motive which results in detrimental environmental impacts. Not only is nature forced to become a victim of capitalist greed and unfettered market competition, man himself also has to bear the 'silent costs' of economic progress (= overuse of human capital) which are estimated to be as large as 10 to 12 percent of gross domestic product (Leipert, 1989).

While not precluding overuse and even exhaustion of natural resources over time, the OLG approach to intertemporal resource economics requires that the belief in capitalism as the sole cause of all resource problems be qualified. The main conclusion from the following model analysis is that intertemporally and intergenerationally inefficient overuse of natural resources can only be attributed to the competitive market system if exclusive property rights to them are lacking.

1.6 Outline of the Book

In closing the first part of this book, Chap. 2 clarifies the different stances towards the compatability (or incompatability) of natural resource use and economic growth.

To avoid unnecessary complication, we focus in Part II on a competitive market economy with economically abundant natural resources. This assumption of resource abundance allows to exclude natural resources from our modeling framework and will thereby serve as a benchmark for our later analysis including natural resources. We start our model analysis by characterizing the intergenerationally efficient market allocation in Chap. 3. The next step, taken in Chap. 4, is to analyze under which conditions an intertemporal market equilibrium is intergenerationally efficient in the short run, while in Chap. 5 we focus on long-run intergenerational efficiency and optimality of the market equilibrium in the steady state.

With this analytical tools at hand, we are then in Part III in a position to broaden our approach by explicitly taking account of natural resources as inputs to production and store of value. Thus, in Chap. 6 the notion of resource regeneration and harvesting are introduced in our basic modeling framework. We again start by characterizing the conditions for short-run intergenerational efficiency before moving on to the conditions for long-run intergenerational efficiency. In Chap. 7, we derive the intertemporal market equilibria under different property rights regimes and discuss whether these equilibria are compatible with short-run and long-run efficiency.

The question of intergenerational equity, namely under which conditions economic growth is sustainable, is addressed in Part IV. In Chap.8 the scope for sustainable economic growth under linear regeneration is dealt with, while in Chap. 9 the feasibility of sustainable development with logistic regeneration is investigated.

In Part V, entitled 'Shocks to Harvest Technology and Natural Regeneration', we integrate costly resource harvest depending inversely on the resource stock (Chap. 10), and analyze in Chap. 11 the consequences of shocks to either the regenerative ability or to harvest costs both on the steady state and during the transition path.

References

Barnett H, Morse C (1963) Scarcity and growth: The economics of natural resource availability. Johns Hopkins University Press, Baltimore

Clark CW (1990) Mathematical Bioeconomics: The Optimal Management of Renewable Resources, 2nd edn. Wiley, NY

Conrad JM (1999) Resource Economics. Cambridge University Press, Cambridge, New York

Conrad JM, Clark CW (1987) Natural Resource Economics: Notes and Problems. Cambridge University Press, Cambridge, New York

Dasgupta PS, Heal G (1979) Economic theory and exhaustible resources. Cambridge University Press, Cambridge

Diamond P (1965) National debt in a neoclassical growth model. American Economic Review 55:1126–1150

Endres A, Querner I (2000) Die Ökonomie natürlicher Resourcen. Kohlhammer, Stuttgart

Farmer K (2000) Intergenerational natural–capital equality in an overlapping–generations model with logistic regeneration. Journal of Economics 72(2):129–152

Farzin Y (2006) Conditions for sustainable optimal economic development. Review of Development Economics 10:518–534

Faustmann (1849) Berechnung des Werthes, welchen Waldboden, sowie noch nicht haubare Holzbestände für die Waldwirthschaft besitzen. Allgemeine Forst- und Jagd-Zeitung 12/1849:441–455

Ghoshray A, Johnson B (2010) Trends in world energy prices. Energy Economics doi:10.1016/j.eneco.2010.01.009

Gray L (1914) Rent under the assumption of exhaustibility. Quarterly Journal of Economics 28:466–489

Hamilton JD (2008) Understanding crude oil prices. Working Paper 14492, NBER

Hanley N, Shogren JF, White B (2007) Environmental Economics in Theory and Practice. Oxford University Press, Cambridge, Oxford

Hayek FA (1945) The use of knowledge in society. American Economic Review 35:519–530

Hayek FA (1973) Law, legislation, and liberty: Rules and order. Routledge & Kegan Paul, London

Hayek FA (1979) Law, legislation, and liberty: the political order of free people. Routledge & Kegan Paul, London

Hotelling H (1931) The economics of exhaustible resources. Journal of Political Economy 39:137–175

Jevons S (1871) The Theory of Political Economy. Macmillan, London
Johannson PO, Mäler KG (1985) The economics of forestry and natural resources. Blackwell, Oxford
Leipert C (1989) Die heimlichen Kosten des Fortschritts: Wie Umweltzerstörung das Wirtschaftswachstum fördert. Suhrkamp, Frankfurt a.M.
Lomborg B (2001) The skeptical environmentalist: Measuring the real state of the world. Cambridge University Press, Cambridge
Malthus TR (1798) An Essay on the Principles of Population. J. Johnston, London
Meadows D, Meadows D, Randers J, Behrens WW (1972) The limits to growth: A report for the Club of Rome's project on the predicament of mankind. Macmillan, London
Meadows D, Meadows D, Randers J (1992) Beyond the limits: Global Collapse or a sustainable future. Earthscan, London
Menger C (1871) Grundsätze der Volkswirtschaftslehre. Braumüller, Vienna
Mourmouras A (1991) Competitive equilibria and sustainable growth in a life–cycle model with natural resources. Scandinavian Journal of Economics 93(4):585–591
Neher P (1990) Natural Resource Economics: Conservation and exploitation. Cambridge University Press, Cambridge, New York
Pareto V (1909) Manual d' économie politique pure. Girard & Briere, Paris
Pearce DW, Turner RK (1990) Economics of natural resources and the environment. Harvester Wheatsheaf, Hemel Hempstead Herfordshire
Perman R, Ma Y, McGilvray J, Common M (2003) Natural Resource and Environmental Economics, 3rd edn. Pearson, Harlow, London, New York
Pigou A (1920,1962) The economics of welfare, 4th edn. Macmillan, London
Repetto R (1985) The Global Possible. Yale University Press, New Haven, CT
Ricardo D (1817) On the Principles of Political Economy and Taxation, reprint, 1926 edn. Everyman, London
Samuelson PA (1958) An exact consumption-loan model of interest with or without the social contrivance of money. Journal of Public Economics 66:467–482
Simon JL (1996) The ultimate resource 2. Princeton University Press, Princeton
Simon JL, Kahn H (1984) The Resourceful Earth. Basil Blackwell, Oxford
Slade ME (1982) Trends in natural resource commodity prices: An analysis of the time domain. Journal of Environmental Economics and Management 9:122–137
Smith A (1776,1976) An inquiry into the nature and causes of the wealth of nations. Oxford University Press, Oxford
Solow RM (1956) A contribution to the theory of economic growth. Quarterly Journal of Economics 70(1):65–94
Ströbele W (1987) Rohstofförkonomik. Vahlen, München
Varian H (1986) Mikroökonomik. Oldenbourg, München
Walras L (1874) Elements of pure economics. Irwin, Homewood
WCED (1987) Our Common Future. Oxford University Press, Oxford

Chapter 2
Economic Growth and Natural Resources

2.1 Introduction

From the previous chapter we know that natural resources can be distinguished in terms of whether they are renewable or not, whether they are used as consumption good or as input to production, whether private property rights are defined or not, and whether they can be reused or not. Depending on these various characteristics, natural resources might either promote or limit economic growth. For instance, new renewable sources of energy might become available and thus enhance growth potential, e.g. based on solar radiation, or sources of non-renewable resources might be used up and thus restrain growth. The purpose of this chapter is to clarify the different stances towards the compatibility or incompatibility of natural resource use and economic growth.

In the next section different views with regard to economic growth and natural resource utilization will be presented. Then, a short digression on the political economics of GDP growth follows. In Sect. 2.4 the feasibility of economic growth under exhaustible fossil resources is considered.

2.2 Economic Growth and the Use of Natural Resources: Differing Views

In terms of 'green philosophy' (e.g. Meadows et al, 1972; Daly, 1990), economic growth is the main cause of environmental degradation and resource exhaustion. In contrast, in mainstream economics economic growth provides the means to reduce environmental pollution and conserve resources for future generations. However, this is only a very rough characterization of the main opposing views concerning economic growth and natural resources. Colby (1990) introduced a much more differentiated description based on five paradigms of environmental management and development. This is reproduced (in a slightly simplified form) in Table 2.1.

K. Farmer and B. Bednar-Friedl, *Intertemporal Resource Economics: An Introduction to the Overlapping Generations Approach*, DOI 10.1007/978-3-642-13229-2_2,

Table 2.1 The different paradigms of environmental and resource management

Paradigm dimension	Frontier Economics	Environmental Protection	Resource Management	Eco-Development	Deep Ecology
Dominant imperative	Progress: infinite growth and prosperity	Tradeoffs: ecology versus economic growth	Sustainability: constraint for green growth	Ecological security: Co-developing humans and nature	Eco-topia: Anti-growth constrained harmony with nature
Human-nature relationship	Extremely strong anthropocentric	Strong anthropocentric	Modified anthropocentric	Ecocentric	Biocentric
Dominant threats	Hunger, poverty, disease,	Health impacts of pollution, endangered species	Resource degradation, poverty, population growth	Ecological uncertainty, global change	Ecosystem collapse, 'unnatural disasters'
Main themes	Open access/free goods: Exploitation of infinite natural resources	Remedial/defensive: 'Legalize Ecology' as economic externality	Global efficiency: 'economize ecology', interdependence	Generative restructuring: 'ecologize social systems', sophisticated symbiosis	Back to nature, 'Biospecies equality', simple symbiosis
Prevalent property regimes	Privatize (neoclassical) or nationalize (marxist) all property	Privatization dominant except some public parks	Global common laws for conservation of oceans, atmosphere, climate, biodiversity	Global and local commons and private property regimes for intergenerational equity	private + common property set aside for preservation
Who pays	Property owners (public at large)	Taxpayers	'polluter pays'	pollution prevention pays: environmental taxes	Avoid costs by foregoing development
Responsibility for development and management	Property owners: private or state	Fragmentation: development decentralized, management central	Integration across multiple governmental levels	Private/public institutional innovations	Largely decentralized but integrated design and management
Environmental management technologies and strategies	Industrial agriculture, high resource intensity, 'free' markets	end-of-pipe technologies, command-and-control market regulation	Impact assessment, conservation of renewable resources, control of population growth	Resilience management, trough-put scale reduction, low-input agriculture	Stability management, low technology, population reduction
Analytic modeling and planning methodologies	Neoclassical or marxist, closed economic systems, no natural production factors	Neoclassical plus willingness to pay for conservation	Neoclassical plus including natural capital	Ecological economics: biophysical open-systems dynamics	Grassroots bio-regional planning, conservation of cultural and biological biodiversity
Fundamental flaws	Creative but mechanistic, no awareness of reliance on ecological balance	Defined by frontier economics in response to deep ecology critic	Downplays social factors, subtly mechanistic, does not handle true uncertainty	May generate false security: magnitudes of change requires new consciousness	Defined in reaction to frontier economics, organic but not creative; reducing population?

Adapted from Colby (1990).

It is not difficult to see that the approach taken in intertemporal resource economics lies somewhere between the paradigms of environmental protection and resource management in Table 2.1. Since economic growth figures prominently within frontier economics and economic development in general it is important to elaborate a little bit more on its role. Table 2.2 contains the arguments for and against economic growth.

Table 2.2 The arguments for and against economic growth

Economic growth is good, because	Economic growth is bad, because
+ more goods are better + technological unemployment remains low + in growing economies, redistribution of income growth easier than of income levels + financing of development aid by highly developed countries easier + lower trade barriers in advanced economies for products from developing countries + environmental protection and resource conservation easier to finance	- without resource saving technological progress, higher resource exploitation - without labor protection, pressure on unqualified workers rises - infinite economic growth exhausts indispensable and non–substitutable natural resources

To get a better understanding of the factors promoting and hampering economic growth in advanced economies we will use a simple analytical framework. By economic growth we mean the annual growth rate of gross domestic product (GDP), denoted by Y_t, which can be expressed formally by:

$$g_t^Y \equiv \frac{Y_{t+1} - Y_t}{Y_t}, \tag{2.1}$$

where t stands for period of time. The most general approach begins by distinguishing those factors which contribute to economic growth. The main economic contributions to GDP and its growth derive from man-made capital stock K, the labor force or working population L and labor productivity τ. Proceeding as above for the definition of the growth rate of GDP, the corresponding growth rates of the capital stock and the working population are:

$$g_t^K \equiv \frac{K_{t+1} - K_t}{K_t}, \tag{2.2}$$

$$g_t^L \equiv \frac{L_{t+1} - L_t}{L_t}, \tag{2.3}$$

while exogenous and time-stationary labor augmenting technological progress (i.e., the growth rate of labor productivity) is denoted by g^τ.

When the economy depends on natural resources, the extraction of non-renewable resources S and harvesting of renewable resources R contribute to output. Since the extraction of the non-renewable resource stock is, by definition, equal to the decrease of its stock, the extraction rate of non-renewable resources in period t is given by:

$$g_t^S \equiv \frac{S_{t+1} - S_t}{S_t}, \tag{2.4}$$

while the per–period harvest rate of the renewable resource stock is denoted by e_t^R.

If moreover the quality of the environment influences output (e.g. the quality of air and water), the state of the environment, denoted by Q, should be considered too. The deterioration rate of environmental quality in period t is then:

$$g_t^Q \equiv \frac{Q_{t+1} - Q_t}{Q_t}. \tag{2.5}$$

Before being able to denote a general identity for GDP growth, let constant production elasticities be denoted by α_K for man made capital, by α_L for labor, by α_S for non-renewable resources, by α_R for renewable resources, and by α_Q for environmental quality. These production elasticities indicate the percentage GDP increase for a one percent increase in respective output. As usual, under constant returns to scale, the total of all production elasticities equals one.

Thus, GDP growth can be described as the product of economic factors (first set of parentheses) and ecological factors (second set of parentheses):

$$g_t^Y \equiv \left(\alpha_K g_t^K + \alpha_L g^L + \alpha_L g^\tau\right)\left(\alpha_S g_t^S + \alpha_R e_t^R + \alpha_Q g_t^Q\right). \tag{2.6}$$

The beauty of this growth identity is that it can be used to explain the following different growth scenarios: (1) economic progress over time, (2) economic crash in the medium to long term, and (3) 'ecological' development, corresponding to a sustainable development path. We will consider each of these scenarios in turn.

Scenario 2.1 (Progress scenario). In an early phase of economic development per capita income is low, natural resources are abundant, and environmental pollution is low due to low economic activity. Formally, this scenario results from the general growth identity when $\alpha_Q = 0$ and $\alpha^S g_t^S + \alpha^R e^R = 1$. Taking this into account, (2.6) becomes:

$$g_t^Y = \alpha_K g_t^K + \alpha_L g^L + \alpha_L g^\tau.$$

In a steady state the GDP growth rate is time-stationary, i.e. $g_t^Y = g^Y$, and GDP and man-made capital grow at the same rate, i.e. $g^Y = g^K$. In other words, the capital/output ratio (capital coefficient) is time-stationary, which accords well with one of the 'stylized facts' of growing advanced economies first explored by Kaldor (1961).

Assuming constant returns to aggregate production, it follows that $\alpha_K = 1 - \alpha_L$. As a consequence, (2.6) becomes eventually:

$$g^Y = g^L + g^\tau. \tag{2.7}$$

Here we encounter the core of neoclassical growth theory of the 1950s and 1960s (see Solow, 1956): In the long-run the GDP growth rate is solely determined by the growth rate of the working population and the growth rate of labor productivity. Ecological and environmental factors do not have any impacts on the long-run GDP growth rate. Hence, neoclassical growth theory is encapsulated by the paradigm of frontier economics in Table 2.1. This scenario is well suited to early phases of economic development and catch-up growth typically found in the post-war period or the period of transition from central planning to market economy.

Scenario 2.2 (Crash scenario). This scenario is characterized by the complete exhaustion of non-renewable resources ($g^S = 0$) and by the destruction of regeneration capacity of renewable resources ($e^R = 0$). It is easy to see that under these assumptions the growth identity 2.6 becomes:

$$g^Y = \left(\alpha_Q g^Q\right)\left(\alpha_K g^K + \alpha_L g^L + \alpha_L g^\tau\right). \tag{2.8}$$

Since $g^Q < 0$, i.e. the environment deteriorates over time, equation 2.8 implies that the steady-state growth rate becomes negative. Obviously, a negative time-stationary growth rate cannot be upheld indefinitely, as sooner or later the economy must collapse.

Roughly speaking, it is this crash scenario which all non–frontier economic paradigms consider to be the long-run outcome of frontier economics. It is therefore no surprise that the paradigm of frontier economics, with neoclassical growth theory as its most prominent proponent, has lost considerable prestige since the 1970s and that the paradigms of environmental protection and resource management have become established subdisciplines of economics.

Scenario 2.3 (Ecological scenario). To derive the ecological scenario from the growth identity 2.6, assume that labor and man-made capital input do not grow, i.e. ($g^K = g^L = 0$), technological progress is completely undermined, i.e. $g^\tau = 0$, non-renewable resources are completely preserved, i.e. $g^S = 0$, and renewable resources are harvested to such an extent that they only regenerate biologically, i.e.

$$e^R = g_t^R \equiv \frac{R_{t+1} - R_t}{R_t}. \tag{2.9}$$

Given these assumptions it is easy to see that the growth identity 2.6 implies zero GDP growth, $g^Y = 0$. However, it is clear that zero growth does not imply a zero production level. Actually, the GDP level is determined by an aggregate production function with time-stationary man-made capital, labor, and harvest of renewable resources as inputs:

$$Y = \mathrm{F}(K, L, e^R). \tag{2.10}$$

While from the point of view of the eco–development and deep–ecology paradigms in Table 2.1, such an ecological scenario is highly desirable, it is highly unlikely, even in advanced countries, that voters would find it acceptable. The reason is that without labor-saving technological progress, i.e. $g^{\tau} = 0$, real wages cannot grow and it stretches the belief to expect workers or their representatives to voluntarily vote for zero-growth of real wages.

How can such questions be resolved in advanced parliamentary democracies? A brief schematic answer is attempted in the next section.

2.3 Political Economics of GDP Growth: A Digression

Consider now a sovereign nation governed by a parliamentary democracy where the citizens have the right to decide by a plebiscite whether the progress scenario or the ecological scenario should be implemented in their country (The crash scenario is not considered as a desirable alternative).

Besides this single electoral decision concerning the two scenarios there is also an ongoing and implicit decision mechanism in the form of consumer purchasing (*consumer sovereignty*). We assume here that consumer goods are sold on competitive markets and that consumers are fully aware of the ecological consequences of their buying decisions.

However, the political and economic preferences either for growth and against resource conservation, or against growth and for resource conservation, vary widely across citizens and consumers. How can the collective preference for one of the two scenarios be derived from individual preferences?

As concerns the collective economic preferences, the answer is clear: through the aggregate demand for an ecologically less or more friendly good. If the latter is larger than the former, producers of the less resource intensive good will win and the suppliers of the other good will lose. Hence, consumers as buyers of goods have 'voted', via their marginal propensity to pay, for the ecological and against the progress scenario.

As known from social choice, a consistent collective political preference is obtained from individual political preferences through preference aggregation provided the individual political preferences are single-peaked. Under this assumption the theory of public choice tells us that the collective political preference order corresponds to the individual political preference of the median voter (Persson and Tabellini, 2002).

Both the economic and the political collective preference is based on the individual economic-ecological preferences or utility function u of a consumer-citizen $h = 1, ..., H$:

$$u^h = u^h(c^h, f^h, Q), \tag{2.11}$$

whereby c^h denotes consumption per capita, f^h is leisure, and Q is environmental quality.

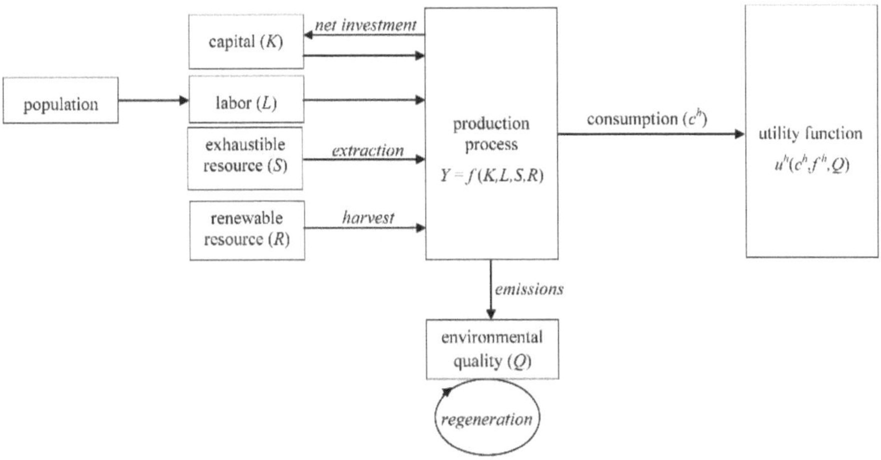

Fig. 2.1 Possibility and desirability of GDP growth

2.4 Economic Growth and Non-Renewable Resources: An Overview

After the publication of the first report to the Club of Rome (Meadows et al, 1972) and the first oil price shock in late 1973, the public in advanced countries quickly became aware of the finiteness of fossil resources such as oil and natural gas. Forecasts of the end of economic growth became common. In reaction to this, eminent economists such as R. Solow and J. Stiglitz initiated a symposium on the possibility of perpetual economic growth under finite fossil resources. The basic insights revealed are summarized below.

A basic question concerning unlimited economic growth with finite fossil resources relates to the conditions under which the economy can grow without being constrained by a lack of fossil resources. Limited non-renewable resources are compatible with unbounded economic growth under three different constellations: (1) when non-renewable resources are not scarce economically, (2) when fossil resources are unnecessary or inessential for production, or (3) after the advent of a backstop technology.

Before being able to discuss these cases in more detail, let us first review the concepts of necessary and essential inputs in production. To do this, we have to dig a little deeper into different forms of aggregate production technologies $F(K,g^S)$. Here F denotes aggregate production as a function of services of man-made capital K, and extraction g^S of the non-renewable resource S.

There are two main types of production functions in the theoretical and empirical literature, Cobb-Douglas (CD) and constant elasticity of substitution (CES). The CD production function has the general structure

$$F(K,g^S) = MK^{\alpha_K}(g^S)^{\alpha_S},$$

where $M > 0$ denotes a constant scaling parameter, and $\alpha_K > 0$ and $\alpha_S > 0$ as before, denote production elasticities. Furthermore, constant returns to scale requires that $\alpha_K + \alpha_S = 1$. The CES production function can be denoted by

$$F(K,e^S) = M\left(\alpha_K K^{-\vartheta} + \alpha_S (g^S)^{-\vartheta}\right)^{-\chi/\vartheta},$$

where $M, \chi, \alpha_K, \alpha_S > 0$, and $-1 < \vartheta \neq 0$. It can be shown that the CD production function is a special case of the CES functional form as ϑ goes to zero at the limit.

A further analytical tool used to characterize aggregate production functions is the elasticity of substitution between K and g^S, denoted by ρ and defined by

$$\rho \equiv \frac{\mathrm{d}(K/g^S)/(K/g^S)}{\mathrm{d}(F_K/F_S)/(F_K/F_S)}.$$

For the elasticity of substitution of the CES production function we get $\rho = 1/(1+\vartheta)$ and hence the elasticity of substitution is constant and does not vary with K/g^S.

With these analytical tools at hand, we are now able to define exactly the necessity and essentialness of non-renewable resources in production.

Definition 2.1. An non-renewable resource is *necessary* if output is zero whenever the quantity of the input of the non-renewable resource is zero, i.e. $F(K,0) = 0$. An non-renewable resource is *essential* if consumption per capita declines over time whenever the quantity of the input of the non-renewable resource is zero, i.e. $g^S = 0 \Rightarrow g_t^{C/L} \leq 0$, whereby C/L denotes consumption per capita.

Clearly, with regard to the CD production function both inputs are necessary. In contrast, it is not easy to see that in the case of CES production technology all inputs are necessary if $\vartheta > 0$, while all inputs are not necessary if $\vartheta < 0$. Looking at the definition of the elasticity of substitution ρ above, it can be seen that no input is necessary where $\rho > 1$, and all inputs are necessary where $\rho < 1$. Thus, we are led to the following proposition.

Proposition 2.1. *If the elasticity of substitution between man-made capital and the input of the non-renewable resource equals one, i.e.* $\rho = 1$ *and the production elasticity of capital is larger than the production elasticity of the resource input, i.e.* $\alpha_K > \alpha_S$, *and if there is no technical progress and no population growth, then the non-renewable resource is not essential, and non-declining consumption per capita is feasible over an indefinite time. This is also the case if the non-renewable resource is not necessary or the elasticity of substitution between capital and the resource is larger than one.*

We are now well prepared to discuss the three possible cases in which limited non-renewable resources are compatible with unlimited economic growth.

Case 2.1 (Non-renewable resources are not scarce economically). This case is nowadays merely of historical interest. While any stock of fossil resource is finite, it may not be scarce since economic activity and the demand for the resource is so low that there is in fact relative resource abundance. As a consequence the price of the resource is zero, or if positive, very low. For example, the very low price of OPEC oil in the 1950s was an indicator that although clearly finite it was not really scarce.

Case 2.2 (Fossil resources are unnecessary or inessential for production). A second constellation for infinite growth with limited non-renewable resources occurs when the natural resource is neither *necessary* nor *essential* for the production of commodities. Intuitively, if man-made capital and the non-renewable resource are highly substitutable, i.e. the elasticity of substitution is larger than one, the declining amount of the resource stock does not lead to a reduction in consumption per capita since the smaller resource input can be easily compensated by a rising amount of man-made capital. In the CD-case the resource input is necessary, but still inessential because due to the relatively higher production elasticity of capital, the decline of the resource input and the associated output reduction is overcompensated by rising capital input. Finally, with a growing population, constant per capita consumption is not possible unless there is sufficiently high technological progress. This even holds when the elasticity of substitution of the CES production function is less than unity (all inputs are necessary), provided the share of output going to the resource is sufficiently low.

Case 2.3 (A backstop technology becomes available). Assume that a so-called backstop technology (Nordhaus, 1973) emerges which represents a perfect substitute for the non-renewable resource. Also in this case, when the resource is not necessary, a constant consumption level per capita can be obtained indefinitely. The world economy turns to the backstop technology (e.g. based on unlimited solar energy within a hydrogen economy) after the price of the non-renewable resource gets so high that with a high input of man-made capital, energy can be provided at high but constant unit cost. As a consequence, the non-renewable resource is not used any longer.

Figure 2.2 depicts the price and quantity paths of an non-renewable resource, where extraction costs are zero, and hence the resource price increases exponentially with a time constant interest rate on financial capital (*Hotelling rule*). The Hotelling rule follows from a simple no-arbitrage argument (Farmer, 1987). Suppose that the owner of an oil well is confronted with the market prices of crude oil in period t and period $t+1$, i.e. p_t and p_{t+1} respectively. The resource owner has to decide whether to extract the resource in period t or in $t+1$. Suppose further that the revenue from the extraction can be deposited in a bank at the period's interest rate r. Under these circumstances the resource owner will extract in period t, if $p_t(1+r) > p_{t+1}$. Otherwise, that is if $p_t(1+r) < p_{t+1}$, he will extract in period $t+1$. The resource owner is indifferent between the alternatives if $p_t(1+r) = p_{t+1}$, and in this case in both periods oil is supplied to the market. Given that the resource demand is larger than zero at all positive resource prices, only indifference is compatible with resource market clearing.

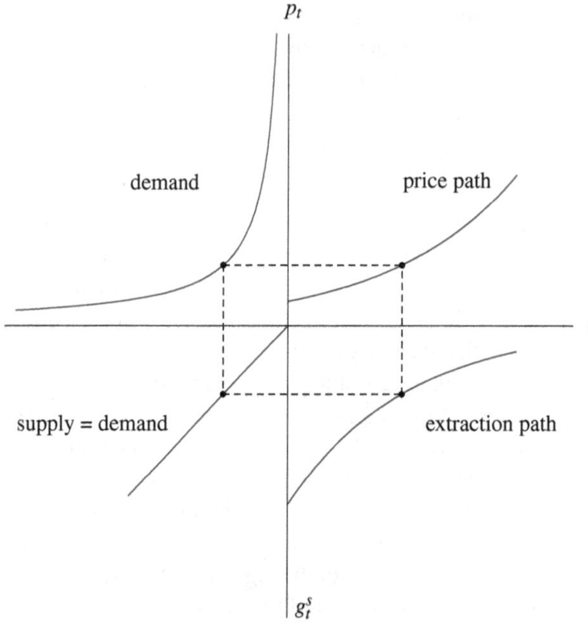

Fig. 2.2 Extraction path for an non-renewable resource over time (Ströbele, 1987, 42)

For an exogenously given interest rate r, $p_t(1+r) = p_{t+1}$ represents a simple difference equation in p_t which can be solved recursively as follows. For $t = 0$ the equation reads as follows: $p_1 = (1+r)p_0$. For $t = 1$ we have: $p_2 = (1+r)p_1 = (1+r)^2 p_0$. Repeating the procedure $T-1$ times, we get eventually: $p_T = (1+r)^T p_0$. If we let the length of any period go to zero, this equation turns to: $p(T) = e^{rT} p_0$. This exponential equation is depicted graphically in the first quadrant of Fig. 2.2 with t on the abscissa and p_t on the vertical axis.

In the second quadrant a unitary elastic resource demand function is drawn with quantity demanded on the abscissa and the resource price on the ordinate. In the third quadrant the 45° line depicts market clearing: resource supply equals demand at all price-quantity combinations. Finally, in the fourth quadrant with quantity extracted on the ordinate and time on the abscissa the extraction path is drawn.

In Fig. 2.3, where only the first quadrant of Fig. 2.2 is depicted, we see, however, that the introduction of constant unit costs c of a backstop technology ends the exponential price path and transforms it into a time-stationary one.

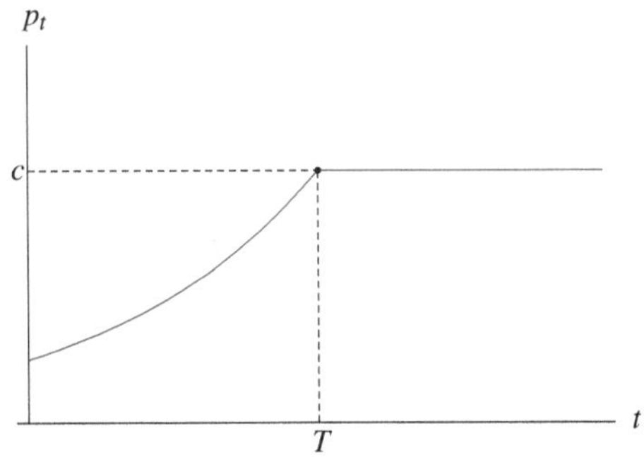

Fig. 2.3 Transition/switch to backstop technology (Ströbele 1987, 46)

2.5 Conclusions

In view of the five paradigms of environmental and resource management, presented at the beginning of the chapter, intertemporal resource economics lies in between the paradigms of environmental protection and resource management. Due to neoclassical growth theory encapsulated by the paradigm of frontier economics GDP growth is determined solely by economic factors. On the other hand, the paradigm of deep ecology will hardly win the majority vote in parliamentary democracies. Indefinite economic growth is compatible with finite exhaustible resources when (i) in stages of early economic development fossil fuels are not scarce, or (ii) man-made capital can be easily substituted for fossil fuels and the later are not important, or (iii) a backstop technology arrives.

References

Colby ME (1990) Environmental management in development. Working Paper 1990:80, World Bank, Washington

Daly H (1990) Toward some operational principles of sustainable development. Ecological Economics 2:1–7

Farmer K (1987) Privatwirtschaftliche Rationalität und volkswirtschaftliches Optimum in der Ökonomik erschöpfbarer Ressourcen. In: Farmer K, Pfeiffer K, Stattegger K (eds) Ökosystemanalyse: Forschungsergebnisse zur Umweltsituation, Leykam, Graz

Kaldor N (1961) Capital accumulation and economic growth. In: Lutz F, Hague D (eds) The Theory of Capital, St. Martin's Press, New York, pp 177–222

Meadows D, Meadows D, Randers J, Behrens WW (1972) The limits to growth: A report for the Club of Rome's project on the predicament of mankind. Macmillan, London

Nordhaus WD (1973) The allocation of energy resources. Brookings Paper on Economic Activity 1973, Brookings Institution, Washington

Persson T, Tabellini G (2002) Political Economics. MIT Press, Cambridge (Mass.)

Solow RM (1956) A contribution to the theory of economic growth. Quarterly Journal of Economics 70(1):65–94

Ströbele W (1987) Rohstoffökonomik. Vahlen, München

Part II
Efficiency and Market Equilibrium under Resource Abundance

Chapter 3
Intergenerational Efficiency in Log-linear Cobb-Douglas OLG Models

3.1 Introduction

Based on the three statements presented in Chap. 2, i.e. (i) exhaustible resources are not scarce, (ii) fossil resources are unnecessary or inessential, and (iii) backstop technology is available, it may appear that the definition and characterization of both intergenerational efficiency and intertemporal market equilibrium are no longer a problem. Unfortunately, this is not in fact the case. Even when both exhaustible and renewable resources are economically abundant, and may therefore be excluded from our modeling framework, an intertemporal market equilibrium may still remain intergenerationally inefficient. We intend to show how such a position may come about in the following pages. By way of illustration we use the most simple version of the Diamond–type (Diamond, 1965) overlapping generations economy with log-linear intertemporal utility and Cobb-Douglas (CD) production function.

In the first subsection we introduce the model economy. Then we define and characterize intergenerationally efficient allocations both analytically and graphically.

3.2 The Log-Linear Cobb-Douglas OLG Economy

We now consider an economy existing over an infinite number of periods $t = 0, 1, \cdots, \infty$. The economy is populated by a large number of households living for two periods and firms operating one period each. Households' decision variables relating to the first period are indexed by the superscript 1 while the second-period decision variables are indexed by the superscript 2. In each time period, L_t young households enter the economy and overlap for one period with the households who entered the economy one period earlier and who therefore constitute the 'old' households in the current period. In the first period of their lifetime younger households work, in the second period they are retired. To keep the analysis as simple as possible, we assume that the number of hours younger households work per period is

K. Farmer and B. Bednar-Friedl, *Intertemporal Resource Economics: An Introduction to the Overlapping Generations Approach*, DOI 10.1007/978-3-642-13229-2_3,

exogenous. Hence, we can without loss of generality specify that each young household works exactly one unit of time. Thus, aggregate labor available for commodity production during period t is equal to L_t.

In each period t, firms produce one homogeneous commodity, denoted by Y_t, which can be either consumed or invested to augment the next period aggregate stock of real capital: K_{t+1}. To produce this commodity, firms employ N_t workers as well as the services of the real capital stock held by older households at the beginning of period t. The production technology is specified by the following CD production function:

$$Y_t = (a_t N_t)^{1-\alpha} (K_t)^{\alpha}, \quad 0 < \alpha < 1, \tag{3.1}$$

where a_t denotes the efficiency of employed labor (productivity), α is the production elasticity of capital (for simplicity we deleted the subscript K) and $1-\alpha$ is the production elasticity of labor. Labor efficiency is not assumed to be time-stationary but evolves over time according to the following rule:

$$a_t = G^{\tau} a_{t-1}, \; G^{\tau} \equiv 1 + g^{\tau}, \; a_{-1} = 1, \tag{3.2}$$

where we assume that the growth factor of labor efficiency (productivity) G^{τ} is identically equal to one plus the exogenously given growth rate of labor efficiency, g^{τ}. Hence, technological progress is modeled as *labor-augmenting*.

We also assume that the working population (number of young households) is growing over time by an exogenously given growth rate of working population, g^L:

$$L_t = G^L L_{t-1}, \; G^L \equiv 1 + g^L, \; L_{-1} = 1. \tag{3.3}$$

The so–called natural growth factor of the economy is then given by

$$\frac{(a_{t+1} L_{t+1})}{(a_t L_t)} = G^L G^{\tau} \equiv G^n. \tag{3.4}$$

Although in each period only the younger generation works, the central planner has not only to take into account the consumption needs of the younger generation (where c_t^1 denotes consumption of a single younger household), but also those of the older generation (c_t^2 denoting consumption of a single older household). However, the aggregate consumption of the younger generation $L_t c_t^1$ plus the aggregate consumption of the older generation $L_{t-1} c_t^2$ cannot be larger than the gross product Y_t plus the non-depreciated capital stock $(1-\delta)K_t$, with δ denoting the exogenous depreciation rate, minus gross investment K_{t+1}:

$$L_t c_t^1 + L_{t-1} c_t^2 \leq Y_t + (1-\delta) K_t - K_{t+1}. \tag{3.5}$$

Rearranging this inequality yields another interpretation of the aggregate consumption-investment constraint:

$$L_t c_t^1 + L_{t-1} c_t^2 + K_{t+1} - K_t \leq Y_t - \delta K_t, \tag{3.6}$$

which says that aggregate consumption plus aggregate net investment cannot be larger than the net product, i.e. gross product minus capital depreciation. Since the CD production function is linearly homogeneous, it is useful to turn the aggregate consumption–investment constraint into per–efficiency–capita terms by dividing the constraint on both sides by $a_t L_t$:

$$\frac{c_t^1}{a_t} + \frac{L_{t-1}c_t^2}{a_t L_t} + \frac{K_{t+1}}{a_{t+1}L_{t+1}} \frac{a_{t+1}L_{t+1}}{a_t L_t} - \frac{K_t}{a_t L_t} \leq \frac{Y_t}{a_t L_t} - \delta \frac{K_t}{a_t L_t}. \tag{3.7}$$

By utilizing the definition of per-efficiency capital intensity $k_t = K_t/(a_t L_t)$ and of the natural growth factor G^n, this inequality can be rewritten as:

$$\frac{c_t^1}{a_t} + \frac{c_t^2}{a_{t-1}G^n} + k_{t+1}G^n - k_t \leq \frac{Y_t}{a_t L_t} - \delta k_t. \tag{3.8}$$

Another constraint the social planner has to respect is that the number of employed younger households cannot be larger than the labor force:

$$N_t \leq L_t, \ \forall t. \tag{3.9}$$

Besides the equality and inequality constraints listed so far, the intertemporal (life-cycle) preferences of younger households with respect to consumption in the working period c_t^1 and in the retirement period c_{t+1}^2 also need to be included as input data in the planners' optimization problem. These preferences are depicted by the following *log-linear* intertemporal utility function:

$$U_t^1 = \ln c_t^1 + \beta \ln c_{t+1}^2, \quad \beta \equiv 1/(1+\theta), \ \theta > 0, \tag{3.10}$$

where θ denotes the time preference rate and β is the time discount factor. $\theta > 0$ means that the younger household prefers present over future consumption. It is also worth noting that the log–linear function is an example of a typical neoclassical utility function which ensures positive but decreasing marginal utility. Notice also that all generations adhere to the same intertemporal utility function.

Given the above description of the model economy, we now turn to the definition and characterization of intergenerationally efficient allocations.

3.3 Intergenerational Efficiency

Before we embark on defining intergenerational efficiency within the model economy introduced in the previous section, we should remind the reader that this notion is not very common in the literature. One exception, Page (1997, 582) notes that intergenerational efficiency is usually defined in the sense of intergenerational Pareto optimality. This simply means that an allocation is intergenerationally Pareto optimal or intergenerationally efficient 'if there exists no other feasible allocation that improves utility of at least one household without reducing utility of any other household of any generation' (Wendner, 2005, 21).

To make more institutional sense of this efficiency notion it can be either seen as an allocative norm established by a long-lived benevolent social planner, or as an equilibrium concept as found in cooperative game theory (see Dasgupta and Heal, 1979, chap. 2). In the latter sense, a feasible allocation across generations is intergenerationally efficient if it cannot be blocked by the grand coalition of all subsequent generations. This cooperative interpretation demands that in each period the younger and the older household meet the producer and all negotiate the period-specific allocation under the assumption that subsequent generations will do the same, or in other words that current generations act on the basis of an implicit intergenerational contract. Thus, whether any of these two interpretations is achievable in reality is still a matter of debate in the literature.

Starting with the analytical definition of intergenerational efficiency with respect to our log-linear CD economy, we first define an employment–investment–consumption allocation A_0^∞ as follows:

$$A_0^\infty = \left\langle \{(N_t, K_{t+1}),\, (c_t^1, c_t^2)\}_{t=0}^{\infty} \,|\, N_t \geq 0,\, K_{t+1} \geq 0,\, c_t^1 \geq 0,\, c_t^2 \geq 0 \right\rangle. \tag{3.11}$$

The next step is to define a feasible employment–investment–consumption allocation. The formal definition reads as follows:

Definition 3.1 (Feasible intergenerational allocation). An employment–investment–consumption allocation A_0^∞ is feasible if for exogenously given K_0, L_{-1},and a_{-1} the allocation fulfills (3.1)-(3.9) $\forall t$.

Now we are able to define an intergenerationally efficient employment–investment–consumption allocation.

Definition 3.2 (Intergenerationally efficient allocation). Consider two feasible allocations $A_0^\infty, A_0^{\infty'}$. Then, the allocation A_0^∞ is intergenerationally efficient if there is no other allocation $A_0^{\infty'}$ for which utility of the initially old generation is greater or equal:

$$U_{-1}^1\left(c_0^{2'}\right) \geq U_{-1}^1\left(c_0^2\right), \tag{3.12}$$

and for which utility of all other generations is strictly greater for some t and at least equal $\forall t$:

$$U_t^1\left(c_t^{1'}, c_{t+1}^{2'}\right) \geq U_t^1\left(c_t^1, c_{t+1}^2\right). \tag{3.13}$$

3.4 First Order Conditions for Short-Run Intergenerational Efficiency

This section is devoted to providing the necessary conditions for intergenerational efficiency which can then be used to identify the intergenerationally efficient employment–investment–consumption allocation defined in the previous section. To derive these first order conditions (FOCs) two preliminary considerations are in order.

First, the definition of intergenerational efficiency given above cannot be directly used to derive these conditions. However, it can be shown (in a static context, see Mas-Colell et al, 1995, 562) that an intergenerationally efficient employment–investment–consumption allocation is equivalent to the solution of the following optimization problem:

$$\max \rightarrow \ln c_{-1}^{1} + \beta \ln c_{0}^{2} \tag{3.14}$$

subject to the following constraints:

$$\ln c_t^1 + \beta \ln c_{t+1}^2 \geq \left(U_t^1\right)^{\circ}, \; t = 0, 1, \cdots \infty, \tag{3.15a}$$

$$\frac{c_t^1}{a_t} + \frac{1}{G^n} \frac{c_t^2}{a_{t-1}} + G^n k_{t+1} \leq (k_t)^{\alpha} + (1-\delta) k_t, \; \forall t, \tag{3.15b}$$

where $\left(U_t^1\right)^{\circ}$ are the exogenously given minimum levels of consumption for all other generations but that born in $t = -1$, and k_0 is given.

In problem (3.14)–(3.15b) we have inserted into the aggregate consumption-investment constraint the CD production function and have assumed that the labor force is fully employed, i.e. $L_t = N_t$. In view of the monotonous utility function of the older household in period 0, it is natural to assume this because otherwise utility could be increased by employing more people and this would contradict utility maximization—or in other words: unemployment is inefficient.

Moreover, notice that the zero–period older household maximizes its retirement utility (its consumption from the working period is historically given) subject to two sets of infinite inequalities. The first set of inequality constraints (3.15a) demands that the life-time utility levels of subsequent younger generations do not fall below certain positive minimal levels $\left(U_t^1\right)^{\circ}$. This protects the vested rights of coming generations. The second set of constraints (3.15b) ensures aggregate consumption–investment levels in all coming periods up to an unknown end of world state.

To circumvent the mathematical problem of an infinite number of constraints we introduce the Lagrangian associated with the optimization problem above. As usual, the Lagrangian turns a maximization problem subject to constraints into an unconstrained maximization problem by adding to the maximand of the original problem the constraints multiplied by the so-called Lagrangian multipliers, denoted here by μ_t^c and ϕ_t^Y. The advantage of this mathematical trick is that the Lagrangian is finite, since in the optimum of the Lagrangian, the terms in squared brackets following the Lagrangian multipliers are all equal to zero.

$$\mathscr{L} = \ln c^1_{-1} + \beta \ln c^2_0 + \sum_{t=0}^{\infty} \mu^c_t \left[\ln c^1_t + \beta \ln c^2_{t+1} - \left(U^1_t\right)^{\circ} \right] + \\ + \sum_{t=0}^{\infty} \phi^Y_t \beta^t \left[(k_t)^{\alpha} + (1-\delta) k_t - \frac{c^1_t}{a_t} - \frac{1}{G^n} \frac{c^2_t}{a_{t-1}} - G^n k_{t+1} \right] \tag{3.16}$$

While having solved a mathematical problem, an economic problem still remains. Diamond (1965) pointed out that as time goes to infinity per–efficiency capital intensity might attain an intergenerationally inefficient level. This is the case when the capital intensity is dynamically inefficient, i.e. when the real return on man-made capital (real interest rate) is less than the natural growth rate. Since dynamic efficiency (a real interest rate larger than or equal to the natural growth rate) is necessary for intergenerational efficiency (see de la Croix and Michel, 2002, 86), it follows immediately that at a dynamically inefficient steady-state per–efficiency capital intensity, intergenerational efficiency cannot hold.

There are several ways to solve this problem. First, one possibility is to restrict the parameter ranges of the CD production function needed to assure dynamic efficiency of steady state equilibria (as time goes to infinity). In particular, Galor and Ryder (1991, 389) show that the steady state equilibria are dynamically efficient if $\alpha \geq 1/2$.

Another possibility is to assume a finite number of periods (generations) over which an intergenerationally efficient allocation is considered. Zilka (1990, 371) opts for this solution and calls the solution 'short-run' (intergenerationally) efficient. While we adopt this terminology we now face a new problem in that we still need to determine the capital intensity and the younger consumption at the end of the planing horizon. Suppose that the last generation which is taken into account enters the economy at time $t = T$. Then, we assume that positive minimal levels of k_{T+1} and c^1_T, namely $\underline{k}$ and $\underline{c}^1$ are exogenously given.

Under these assumptions a (short-run) intergenerationally efficient employment–investment–consumption allocation results from the solution of the following optimization problem:

$$\max \rightarrow \ln c^1_{-1} + \beta \ln c^2_0 \tag{3.17}$$

subject to the following constraints:

$$\ln c^1_t + \beta \ln c^2_{t+1} \geq \left(U^1_t\right)^{\circ}, \; t = 0, 1, ..., T-1, \tag{3.18a}$$

$$\frac{c^1_t}{a_t} + \frac{1}{G^n} \frac{c^2_t}{a_{t-1}} + G^n k_{t+1} \leq (k_t)^{\alpha} + (1-\delta) k_t, \; t = 0, 1, ..., T, \tag{3.18b}$$

$$k_{T+1} \geq \underline{k}, \tag{3.18c}$$

$$c^1_T \geq \underline{c}^1. \tag{3.18d}$$

The corresponding Lagrangian reads as follows:

$$\begin{aligned}\mathcal{L} = {} & \ln c_{-1}^1 + \beta \ln c_0^2 + \sum_{t=0}^{T-1} \mu_t^c \left[\ln c_t^1 + \beta \ln c_{t+1}^2 - \left(U_t^1\right)^\circ \right] + \\ & + \sum_{t=0}^{T} \phi_t^Y \beta^t \left[(k_t)^\alpha + (1-\delta) k_t - \frac{c_t^1}{a_t} - \frac{1}{G^n} \frac{c_t^2}{a_{t-1}} - G^n k_{t+1} \right] + \\ & + \phi_{T+1}^K \beta^{T+1} [\underline{k} - k_{T+1}] + \mu_T^c \beta^T [\underline{c}^1 - c_T^1].\end{aligned} \tag{3.19}$$

The first-order conditions for a saddle-point solution of this Lagrangian are as follows, where superscript $\circ$ denotes that the solution is short-run intergenerationally efficient:

$$\frac{\partial \mathcal{L}}{\partial c_0^2} = \frac{\beta}{\left(c_0^2\right)^\circ} - \frac{\left(\phi_0^Y\right)^\circ}{a_{-1} G^n} = 0, \tag{3.20a}$$

$$\frac{\partial \mathcal{L}}{\partial c_t^1} = \frac{\left(\mu_t^c\right)^\circ}{\left(c_t^1\right)^\circ} - \frac{\left(\phi_t^Y\right)^\circ \beta^t}{a_t} = 0, \ \forall t, \tag{3.20b}$$

$$\frac{\partial \mathcal{L}}{\partial c_{t+1}^2} = \frac{\left(\mu_t^c\right)^\circ \beta}{\left(c_{t+1}^2\right)^\circ} - \frac{\left(\phi_{t+1}^Y\right)^\circ \beta^{t+1}}{G^n a_t} = 0, \ \forall t, \tag{3.20c}$$

$$\begin{aligned}\frac{\partial \mathcal{L}}{\partial k_{t+1}} = {} & \left(\phi_{t+1}^Y\right)^\circ \beta^{t+1} \alpha \left[(k_{t+1})^\circ\right]^{\alpha-1} + \left(\phi_{t+1}^Y\right)^\circ \beta^{t+1} (1-\delta) - \\ & - G^n \left(\phi_t^Y\right)^\circ \beta^t = 0, \forall t,\end{aligned} \tag{3.20d}$$

$$\frac{\partial \mathcal{L}}{\partial \mu_t^c} = \ln \left(c_t^1\right)^\circ + \beta \ln \left(c_{t+1}^2\right)^\circ - \left(U_t^1\right)^\circ = 0, \ t = 0, \dots, T-1, \tag{3.20e}$$

$$\frac{\partial \mathcal{L}}{\partial \phi_t^Y} = \left[(k_t)^\circ\right]^\alpha + (1-\delta)(k_t)^\circ - \frac{\left(c_t^1\right)^\circ}{a_t} - \left(\frac{1}{G^n}\right) \frac{\left(c_t^2\right)^\circ}{a_{t-1}} - G^n (k_{t+1})^\circ, \ \forall t, \tag{3.20f}$$

$$\frac{\partial \mathcal{L}}{\partial \phi_{T+1}^K} = \underline{k} - (k_{T+1})^\circ = 0, \tag{3.20g}$$

$$\frac{\partial \mathcal{L}}{\partial \mu_{T+1}^c} = \underline{c}^1 - \left(c_T^1\right)^\circ = 0. \tag{3.20h}$$

Combining (3.20b)–(3.20d) and eliminating the Lagrange multipliers yields an intertemporally optimal consumption pattern for the younger generation alive in period t:

$$\frac{\left(c_{t+1}^2\right)^\circ}{\beta \left(c_t^1\right)^\circ} = \alpha \left[(k_{t+1})^\circ\right]^{\alpha-1} + 1 - \delta, \quad \forall t. \tag{3.21}$$

The left hand side of (3.21) shows the intertemporal marginal rate of substitution between working-period and retirement consumption. The right hand side of (3.21) gives the gross return on real capital invested in period t which equals gross marginal product of capital at the beginning of period $t+1$ plus the non-depreciated portion of one unit invested capital. Thus, short-run intergenerational efficiency demands that the intertemporal marginal rate of substitution between future and present consumption equals the gross real return factor on one unconsumed commodity unit.

In maximizing the utility of zero-period older generation, we need to ensure that no subsequent younger generation suffers a utility loss:

$$\ln\left(c_t^1\right)^\circ + \beta \ln\left(c_{t+1}^2\right)^\circ = \left(U_t^1\right)^\circ, \quad t = 0, \ldots, T-1. \tag{3.22}$$

Furthermore, intergenerational efficiency demands that the aggregate consumption–investment constraint is binding:

$$[(k_t)^\circ]^\alpha + (1-\delta)(k_t)^\circ = \frac{\left(c_t^1\right)^\circ}{a_t} + \frac{1}{G^n}\frac{\left(c_t^2\right)^\circ}{a_{t-1}} + G^n\left(k_{t+1}\right)^\circ, \quad \forall t. \tag{3.23}$$

The terminal conditions require that $k_{T+1} = \underline{k}$ and $c_T^1 = \underline{c}^1$.

Finally, (3.20a) needs to hold for $t = 0$:

$$\frac{\beta}{\left(c_0^2\right)^\circ} = \frac{(\mu_0)^\circ}{a_{-1}G^n}. \tag{3.24}$$

This optimality condition ensures that the older generation in period zero is ready to accept the consumption level c_0^2, which follows from the aggregate consumption-investment constraint in period zero. In terms of achieving intergenerational efficiency, this the only crucial intergenerational efficiency condition.

3.5 Graphical Illustration of FOCs for Short-Run Intergenerationally Efficient Allocation

To be able to illustrate the FOCs for short-run intergenerational efficiency we set $T = 1$, and we assume for the sake of simplicity that $\delta = 1$ and $G^L = G^\tau = G^n = 1 = a_{-1} = L_{-1}$. Consider now for $t = 0, 1$ the aggregate consumption-investment constraint. For exogenously given $\underline{k}_0$ and due to the terminal conditions $\left(c_1^1\right)^\circ = \underline{c}^1$ and $(k_2)^\circ = \underline{k}$, we get for the aggregate consumption-investment constraint for periods zero and one:

$$(c_0^1)^\circ + (c_0^2)^\circ + (k_1)^\circ = (\underline{k}_0)^\alpha, \tag{3.25}$$

$$\underline{c}^1 + (c_1^2)^\circ + \underline{k} = ((k_1)^\circ)^\alpha. \tag{3.26}$$

Note that furthermore $(c_0^2)^\circ$ is given from the perspective of the young generation in period 0. Then, (3.25) can be solved for $(k_1)^\circ$ and the result inserted into (3.26) to yield a unique relationship between the present and the retirement consumption of a zero-period younger household:

$$(c_1^2)^\circ = \left[(\underline{k}_0)^\alpha - (c_0^1)^\circ - (c_0^2)^\circ\right]^\alpha - \underline{c}^1 - \underline{k}. \tag{3.27}$$

This equation, for the exogenously given parameters above, can be termed the intertemporal transformation curve and is depicted in Fig. 3.1. Its slope, the intertemporal marginal rate of transformation (MRT), is determined as follows:

$$MRT \equiv -\frac{\mathrm{d}c_1^2}{\mathrm{d}c_0^1} = \alpha\,(k_1)^{\alpha-1}.$$

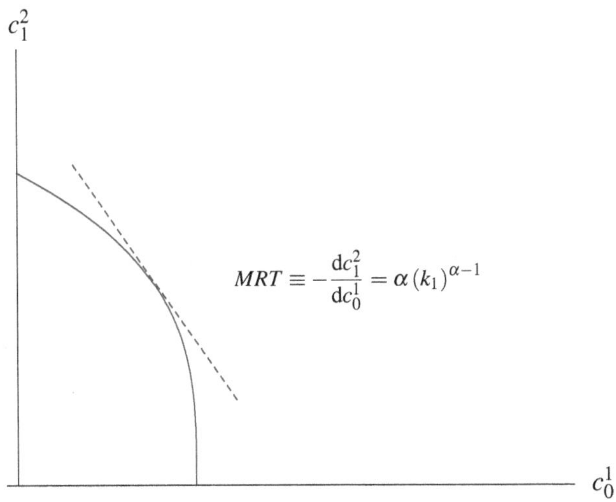

Fig. 3.1 Intertemporal transformation curve and its slope

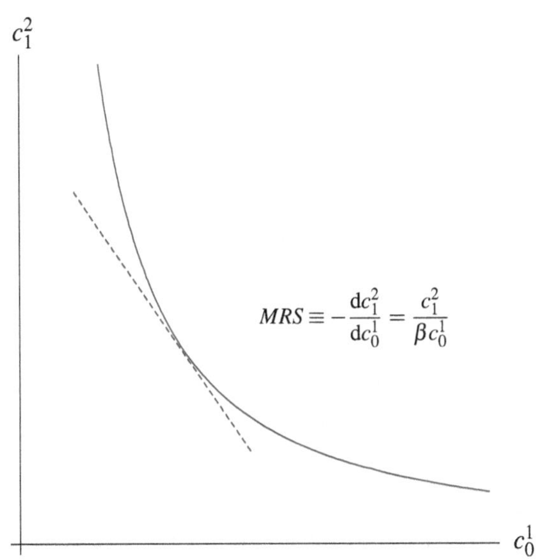

Fig. 3.2 Intertemporal indifference curve for $u\left(c_0^1, c_1^2\right) = \left(\bar{U}_0^1\right)^{\circ}$ and its slope

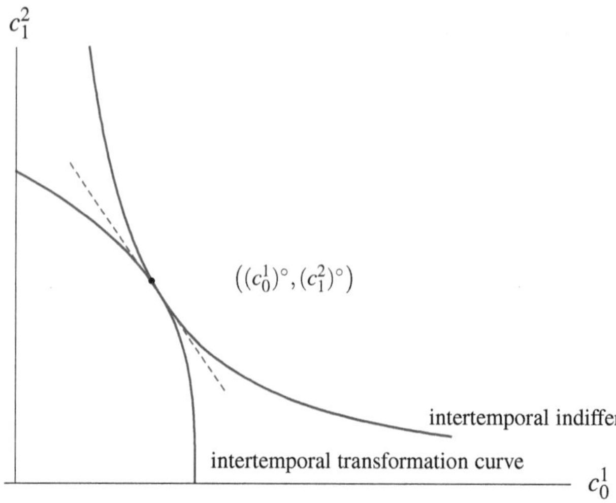

Fig. 3.3 Short-run intergenerationally efficient allocation $\left((c_0^1)^\circ, (c_1^2)^\circ\right)$ where $MRS = MRT$ holds

In addition, we also need the intertemporal indifference curve which is also depicted in Fig. 3.2. This curve can be derived by setting the intertemporal utility function of zero-period younger generation equal to the fixed level $\bar{U}_0^1$ and solving the result for c_1^2:

$$\ln c_0^1 + \beta \ln c_1^2 = \left(\bar{U}_0^1\right)^\circ \Leftrightarrow c_0^1 \cdot \left(c_1^2\right)^\beta = \left(\bar{U}_0^1\right)^\circ \Leftrightarrow c_1^2 = \left(\frac{\left(\bar{U}_0^1\right)^\circ}{c_0^1}\right)^{\frac{1}{\beta}}. \tag{3.28}$$

The slope of this indifference curve, termed the intertemporal marginal rate of substitution (MRS), is calculated as follows:

$$MRS \equiv -\frac{dc_1^2}{dc_0^1} = \frac{c_1^2}{\beta c_0^1}.$$

Equating the intertemporal marginal rate of transformation with the intertemporal marginal rate of substitution (as illustrated in Fig. 3.3) yields

$$\frac{(c_1^2)^\circ}{\beta (c_0^1)^\circ} = \alpha \left((k_1)^\circ\right)^{\alpha-1}, \tag{3.29}$$

which together with equation (3.25) gives two equations to determine the two unknowns $(c_0^1)^\circ$ and $(c_1^2)^\circ$. Inserting the solutions for these variables into equations (3.25) and (3.26) leaves us with two equations to determine the remaining variables $(k_1)^\circ$ and $(c_0^2)^\circ$. This shows that consumption of a zero-period older household c_0^2 is not really a parameter, as temporarily assumed above, but an endogenous variable determined by the whole set of conditions for short-run intergenerational efficiency.

3.6 Conclusions

This chapter provided a precise definition and an exact characterization of short-run intergenerationally efficient consumption allocations within Diamond's 1965 log-linear OLG economy. After figuring out the analytical and substantial difficulties in defining intertemporal efficiency, Zilka's (1990) definition of short-run intergenerational efficiency is adopted, and is characterized by first order conditions. To illustrate the FOCs graphically, a simplified version of a two-period economy is used. It turns out that short-run intergenerational efficiency essentially boils down to the equality of the intertemporal marginal rate of substitution and the intertemporal marginal rate of transformation.

This chapter was devoted to efficiency issues. In the next chapter we introduce the market analogue to the planner problem, as well as define and characterize intertemporal market equilibria.

References

de la Croix D, Michel P (2002) A Theory of Economic Growth. Dynamics and Policy in Overlapping Generations. Cambridge University Press, Cambridge, New York

Dasgupta PS, Heal G (1979) Economic theory and exhaustible resources. Cambridge University Press, Cambridge

Diamond P (1965) National debt in a neoclassical growth model. American Economic Review 55:1126–1150

Galor O, Ryder HR (1991) Dynamic efficiency of steady-state equilibria in an overlapping generations model with productive capital. Economic Letters 35:385–390

Mas-Colell A, Whinston MD, Green JR (1995) Microeconomic Theory. Oxford University Press, Oxford

Page T (1997) On the problem of achieving efficiency and equity, intergenerationally. Land Economics 73:580–596

Wendner R (2005) Frames of reference, the environment, and efficient taxation. Economics of Governance 6:13–31

Zilka I (1990) Dynamic efficiency in overlapping generations models with stochastic production. Journal of Economic Theory 52:364–379

Chapter 4
Intertemporal Market Equilibrium and Short-Run Intergenerational Efficiency

4.1 Introduction

The main finding of Chap. 3 was that short-run intergenerational efficiency reduces basically to intertemporal efficiency, or in other words, the specific demographic assumptions of overlapping generations do not generate results essentially different from those of infinitely lived agent models described in the seminal Ramsey (1928) model.

However, the notion of intergenerational or intertemporal efficiency is largely devoid of any realistic institutional structure, in particular with respect to Hayek's 'Great Society' 'which does not presuppose the pursuit of specific goals but which enables individuals who do not know one another intensively and pursue competing objectives to utilize the specific knowledge of others individuals to pursue their own objectives.' (Hayek, 1973, 216) As Hayek (1988) convincingly demonstrated, only a competitive market economy is able to cope with the problem of acquiring and utilizing knowledge in a world comprising billions of decentralized agents. Translated to the present context, the question now is whether the benchmark of short-run intergenerational efficiency can be realized by a competitive market economy with overlapping generations.

To answer this analytically we need an abstract model of a competitive market economy evolving over time. As Farmer and Stadler (2005, 70) argue there are two theoretical approaches to a competitive market economy: the neoclassical general equilibrium model of perfect competition developed by Walras (1874) and Pareto (1909), and later analytically refined by Arrow and Debreu (1954), and the neo-Austrian theories of market co-ordination and entrepreneurial innovation developed by Hayek (1963) and Kirzner (1973). Due to its greater analytical tractability, in the following we utilize the neoclassical approach.

Thus, in the next section the highly abstract institutional framework of the neo-classical market model is described. Then, the two-layer structure of neoclassical general equilibrium models, namely the individual optimization problems on the one hand and the market clearing conditions on the other, is applied. In Sect. 4.5

K. Farmer and B. Bednar-Friedl, *Intertemporal Resource Economics: An Introduction to the Overlapping Generations Approach*, DOI 10.1007/978-3-642-13229-2_4,

the intertemporal equilibrium dynamics is derived from the FOCs for intertemporal utility and one-period profit maximization and market clearing conditions for all periods. Finally, a version of the first theorem of welfare economics is applied to the log-linear CD OLG economy.

4.2 The Institutional Framework

The existence of markets presupposes the definition and implementation of exclusive private property rights with respect to resources and commodities. We thus assume here that man-made capital at the beginning of period zero is privately owned by the older household. It sells the services of the capital stock to producing firms and sells any remaining capital stock (after depreciation) to younger household. The younger household owns its human capital and sells its services to producing firms. The producing firms own the commodities produced and sell them to the older and to the younger household.

The resources and commodities the agents own do not in general conform to the most preferred combination of resources and commodities. To attain the latter, there is scope for markets, where production factors and produced commodities are traded. Following Walras (1874), we assume that all markets are highly organized and that traders are perfectly informed about the current prices at which factors and produced goods are exchanged.

Moreover, we assume that a single seller or buyer on any factor or product market cannot influence the market price by his or her decision, i.e. all agents in all markets are price takers. Even if markets have not yet cleared it is assumed that all traders base their decision on signals from an imaginary auctioneer such that a uniform price exists for all items. Finally, under market disequilibrium no quantity rationing arises—a fact which can be also attributed to the activities of the auctioneer.

As a consequence, the centralized and cooperative decision problem to determine a short-run intergenerationally and intertemporally efficient allocation is turned into a series of decentralized and non-cooperative (individual) decision problems. This allows for huge savings with respect to the costs of revealing and gathering private information. It also has the consequence that the solution of individual optimization problems can be separated from the problem of finding an allocation compatible with the aggregate consumption-investment constraint. However, there is still the need for a coordination mechanism which processes and correctly transmits the private information of one market participant to the other market participants and makes the isolated individual decisions compatible in the aggregate. The market, a system of flexible and interdependent equilibrium prices, fulfils this socially beneficial function exactly.

In our highly simplified model economy there are only two (relative) prices available for the coordination of individual decisions: these are the real wage rate w_t (nominal wage expressed in terms of units of the produced commodity) and the real price of capital services r_t which is equal to the real interest rate on savings i_t plus the depreciation rate δ. Clearly, the relative price of the produced commodity is unity. To simplify a little we assume that the depreciation rate is also unity: $\delta = 1$.

4.3 Individual Optimization Problems

As in the previous chapter, our model economy is characterized by two types of households, young and old, and one type of production in every period. Every household born in period $t = 0, 1, \ldots$ lives for two periods, a working period when young and a retirement period when old. An old household is also alive in the initial period ($t = 0$).

Starting with the optimization problem of this old household at the start of the economy, we assume that it maximizes its lifetime utility subject to its retirement budget constraint since the consumption of the previous period c^1_{-1} is historically given:

$$\max \ln c^1_{-1} + \beta \ln c^2_0 \quad \text{subject to:} \quad c^2_0 = r_0 \frac{K_0}{L_{-1}}. \tag{4.1}$$

The optimization problem of the younger household in any period t reads as follows:

$$\max \ln c^1_t + \beta \ln c^2_{t+1} \tag{4.2}$$

subject to:

$$c^1_t + s^1_t = w_t, \tag{4.3a}$$

$$c^2_{t+1} = (1 + r_{t+1} - 1)\, s^1_t = (1 + i_{t+1})\, s^1_t. \tag{4.3b}$$

The constraint (4.3a) during working period t demands that consumption c^1_t and savings s^1_t of the younger household are equal to the real wage rate w_t, while the retirement period budget constraint (4.3b) requires that consumption c^2_{t+1} in the retirement period equals one plus the real interest rate on savings, denoted by i_{t+1}.

The typical producer maximizes real profit. This equals the difference between the quantity of units sold (=produced) of the commodity and the real factor costs, these consisting of real labor and capital costs in any period t:

$$\max \pi_t = Y_t - w_t N_t - r_t K_t \tag{4.4}$$

subject to:

$$Y_t = (a_t N_t)^{1-\alpha} (K_t)^{\alpha}, \tag{4.5a}$$

$$N_t \geq 0,\ K_t \geq 0. \tag{4.5b}$$

The optimization problem of the older household at the start of the economy has the following trivial solution since there is only one decision variable:

$$L_{-1}\left(c_0^2\right) = r_0 K_0. \tag{4.6}$$

To derive the first order condition (FOC) of the younger generation born in any period $t = 0, 1, ...,$ the retirement period budget constraint is solved for s_t^1 and the result then inserted into the working period budget constraint which yields the so-called *intertemporal* budget constraint:

$$c_t^1 + \frac{c_{t+1}^2}{r_{t+1}} = w_t. \tag{4.7}$$

Then, we maximize (4.2) subject to (4.7), set up the corresponding Lagrangian $\mathscr{L}$, differentiate it with respect to c_t^1 and c_{t+1}^2, and after eliminating the Lagrangian multiplier we obtain the following marginal FOC:

$$\frac{1}{\beta}\frac{c_{t+1}^2}{c_t^1} = r_{t+1}. \tag{4.8}$$

The interpretation of (4.8) is straightforward: For an intertemporal optimal bundle of present and future consumption the intertemporal marginal rate of substitution (MRS) between present and future consumption (which equals the time preference factor times the ratio of future to present consumption) has to be equal to the real price of capital services, in period $t+1$. Figure 4.1 illustrates this intertemporal utility maximizing solution showing present consumption on the abscissa and future consumption on the vertical axis.

On utilizing both the intertemporal budget constraint (4.7) and the marginal FOC (4.8) it is easy to calculate the levels of intertemporal utility maximizing consumption and savings for the working period:

$$c_t^1 = \frac{w_t}{(1+\beta)}, \tag{4.9}$$

$$s_t^1 = \frac{\beta}{(1+\beta)} w_t. \tag{4.10}$$

The FOCs for intratemporal profit maximization are easily derived as follows. Insert (4.5a) into the profit function (4.4), differentiate the profit function with respect to K_t and N_t and set the results equal to zero. Using again the per–efficiency capita notation where $k_t = K_t/(a_t N_t)$, the first profit maximization condition requires that the marginal productivity of labor is equal to the real wage rate:

$$(1-\alpha)(k_t)^{\alpha} a_t = w_t, \tag{4.11}$$

and the second that the marginal productivity of capital is equal to the rental price of capital:

$$\alpha (k_t)^{\alpha-1} = r_t. \tag{4.12}$$

Finally, due to the assumption of a constant returns CD production function, maximum profit is zero or in other words, production equals real factor costs:

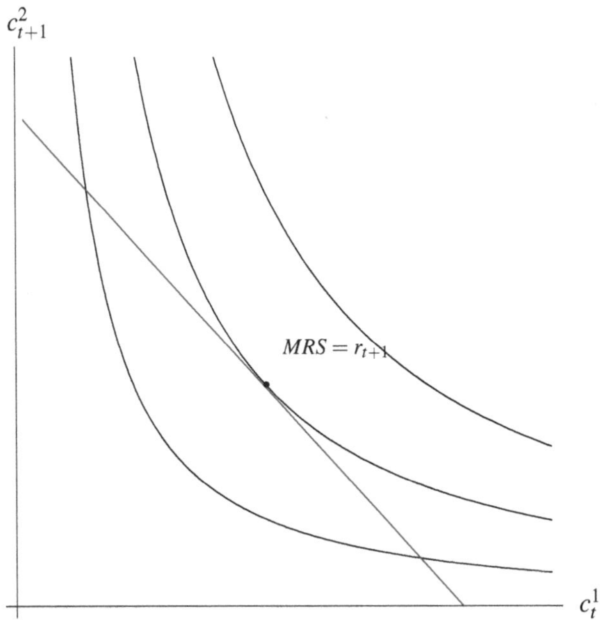

Fig. 4.1 Intertemporal utility maximizing problem

$$a_t N_t k_t^{\alpha} = w_t N_t + r_t a_t N_t k_t. \tag{4.13}$$

4.4 Market Clearing Conditions

There are three perfectly competitive markets in each period in the model economy. Clearing of the commodity market in each period t demands:

$$a_t N_t k_t^{\alpha} = L_t c_t^1 + L_{t-1} c_t^2 + K_{t+1}. \tag{4.14}$$

Labor market clearing in each period reads as follows:

$$N_t = L_t. \tag{4.15}$$

Clearing of the market for capital services demands:

$$a_t N_t k_t = K_t. \tag{4.16}$$

However, there are only two relative prices as stated above. Hence, with three equations and only two endogenous variables in each period a logical inconsistency seems to emerge. On employing Walras law, formulated in the following proposition, the inconsistency disappears.

Proposition 4.1 (Walras law). *If in each market period t* (4.15) *and* (4.16) *hold,* (4.14) *also holds.*

Proof. Add up the current-period budget constraints of the younger and the older household, and you get: $L_t c_t^1 + L_{t-1} c_{t-1}^2 + L_t s_t^1 = L_t w_t + r_t K_t$. Since on account of the assumptions that the markets for labor and capital services clear, we know that $L_t = N_t$ and $K_t = a_t N_t k_t$. Hence, the fact that maximal profits are zero, i.e. $a_t N_t k_t^{\alpha} = w_t N_t + r_t a_t N_t k_t$, implies that $L_t c_t^1 + L_{t-1} c_{t-1}^2 + L_t s_t^1 = a_t N_t k_t^{\alpha}$ holds. Clearly, this equation is equivalent to commodity market clearing if $L_t s_t^1 = K_{t+1}$. □

The following proposition shows that the condition $L_t s_t^1 = K_{t+1}$ is a consequence of the market clearing conditions and that therefore aggregate savings of younger households are equal to the capital stock in the following period.

Proposition 4.2. *Clearing of the labor, capital service and commodity market in each period implies*

$$K_{t+1} = L_t s_t^1. \tag{4.17}$$

Proof. Restating (4.14), taking account of (4.13) and (4.3) yields:

$$w_t N_t + r_t a_t N_t k_t = w_t L_t - L_t s_t^1 + L_{t-1} r_t s_{t-1}^1 + K_{t+1}. \tag{4.18}$$

Since (4.15) requires that $N_t = L_t$ and since (4.16) requires that $a_t N_t k_t = K_t$, it must be the case that $K_{t+1} = L_t s_t^1, \forall t$ is true. □

Note that this savings-investment equality does not presuppose a depreciation rate of one. It also holds for smaller depreciation rates. Moreover, (4.17) forms the basis for the dynamics of the intertemporal equilibrium derived in the next section.

4.5 Intertemporal Equilibrium Dynamics

The derivation of the intertemporal equilibrium dynamics starts with the savings-investment equality of the previous section. Dividing (4.17) on both sides by $a_t L_t = a_t N_t = A_t$ gives:

$$\frac{K_{t+1}}{A_t} = \frac{s_t^1}{a_t}. \tag{4.19}$$

Expansion of the left-hand side by A_{t+1} gives:

$$\frac{K_{t+1}}{A_{t+1}} \frac{A_{t+1}}{A_t} = k_{t+1} \frac{A_{t+1}}{A_t} = \frac{s_t^1}{a_t}. \tag{4.20}$$

Acknowledging that $A_{t+1}/A_t = (a_{t+1} L_{t+1})(a_t L_t) = G^{\tau} G^{L} \equiv G^{n} \approx 1 + g^{n}$ in (4.20) and using the resulting expression in (4.17) we receive:

$$k_{t+1} = \frac{s_t^1}{G^n a_t}. \tag{4.21}$$

Inserting (4.11) into (4.10), gives the following expression for s_t/a_t:

$$\frac{s_t}{a_t} = \frac{\beta(1-\alpha)(k_t)^\alpha}{(1+\beta)}. \tag{4.22}$$

Insertion of (4.22) into (4.21) gives:

$$k_{t+1}G^n = \frac{\beta(1-\alpha)(k_t)^\alpha}{(1+\beta)}. \tag{4.23}$$

After defining $\sigma \equiv \beta(1-\alpha)/(1+\beta)$, the fundamental equation of motion of per-efficiency capital intensity within the OLG model with natural resource abundance results:

$$k_{t+1} = \frac{\sigma}{G^n}(k_t)^\alpha, \text{ for } t = 0, \dots \text{ and given } \quad k_0 = \frac{K_0}{a_0 L_0}. \tag{4.24}$$

Figure 4.2 illustrates the fundamental equation of motion of per–efficiency capital intensity, and determines per–efficiency capital intensity in period $t+1$ as a function of per–efficiency capital intensity in period t. Due to the functional form of the production function (CD technology with constant returns to scale), the larger the capital stock in period t, the smaller the rate of growth of capital (the slope of the function) becomes.

Based on the fundamental equation of motion, it is straightforward to derive the paths of the factor prices and consumption allocation. From the marginal productivity conditions (4.11) and (4.12) the paths of intertemporal equilibrium real wage and real capital service price follow:

$$w_t = a_t(1-\alpha)(k_t)^\alpha \; t = 0, \dots, \tag{4.25}$$

$$r_t = \alpha(k_t)^{\alpha-1}, \; t = 0, \dots. \tag{4.26}$$

With k_t, w_t and r_t determined for $t = 0, \dots$, consumption allocation $\left(c_0^2, c_t^1, c_{t+1}^2\right)$ and capital intensities k_{t+1} in the intertemporal market equilibrium are also determined for $t = 0, \dots$. Having derived the intertemporal market allocation, it is natural to ask whether this allocation is intergenerationally efficient in the short run. The answer is given in the next section.

4.6 Short-Run Intergenerational Efficiency of the Intertemporal Market Equilibrium

It might be thought that the market allocation described in the previous section need not necessarily satisfy the FOCs for short-run intergenerational efficiency as defined

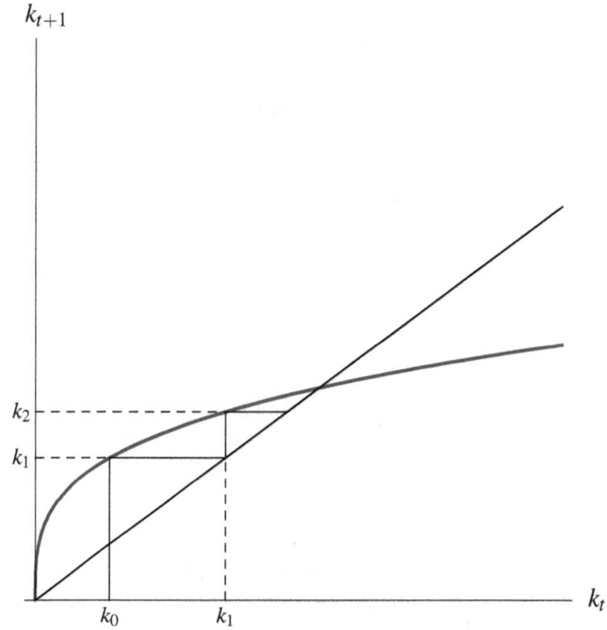

Fig. 4.2 The dynamics of per–efficiency capital intensity

in Chap. 3. However, we will show in this section that the consumption allocation following from an intertemporal market equilibrium over a finite number of market periods is in fact short-run intergenerationally efficient.

Proposition 4.3 (First theorem of welfare economics). *The consumption allocation* $\left\{ \left(c_t^1, c_t^2,\right) \middle| t = 0, 1, ..., T \right\}$ *and the time path of capital intensities* $\langle k_{t+1}, t = 0, 1, ..., T \rangle$ *along an intertemporal market equilibrium are short-run intergenerationally efficient.*

Before embarking on the 'proof' of the proposition (strictly speaking it is mere a demonstration of the equivalence of the FOCs for intertemporal market equilibrium and for short-run intergenerational efficiency) it seems to be useful to outline the thinking behind the proof. The idea is to start with the individual FOCs and market clearing conditions for an intertemporal market equilibrium. The next step is to assume that the quantity variables determined by the FOCs for intertemporal market equilibrium are equal to the quantity variables following from the FOCs for short-run intergenerational efficiency. Finally, we have to show that by simple manipulation of the FOCs for intertemporal market equilibrium, those for short-run intergenerational efficiency follow.

Proof. Consider first the optimality condition which determines consumption demand of the older household in the starting period $t = 0$ in the following, slightly rewritten form (i.e. using (3.3) and (4.24) in (4.6)):

$$c_0^2 = r_0 k_0 G^L a_0. \tag{4.27}$$

To simplify the exposition without compromising the generality of the argument, we rewrite the intertemporal budget constraint (4.7) and the FOC of the younger household (4.3) for $T = 1$:

$$c_0^1 + \frac{c_1^2}{r_1} = w_0, \tag{4.28}$$

$$(1+\theta)\left[\frac{c_1^2}{c_0^1}\right] = r_1. \tag{4.29}$$

For periods $t = 0, 1$, the FOCs for profit maximization can be rewritten as:

$$(1-\alpha)\left(k_t\right)^{\alpha} a_t = w_t,\ t = 0,1, \tag{4.30}$$

$$r_t = \alpha\left(k_t\right)^{\alpha-1},\ t = 0,1. \tag{4.31}$$

The next step is to divide the zero-period market clearing condition for the commodity market (i.e. (4.14) for $t = 0$) on both sides by $a_0 L_0$ and expand it for $a_1 L_1$. This yields:

$$L_0 c_0^1 + L_{-1} c_0^2 = a_0 N_0 \left(k_0\right)^{\alpha} - K_1 \left| \frac{1}{a_1 L_1} \frac{a_1 L_1}{a_0 L_0} \right. . \tag{4.32}$$

Acknowledging furthermore (3.2), (4.24), (3.3), and $G^{\tau} G^L \equiv G^n$ in (4.32) gives

$$\frac{c_0^1}{a_0} + \frac{c_0^2}{a_0 G^L} = \left(k_0\right)^{\alpha} - k_1 G^n. \tag{4.33}$$

Again, the first-period commodity market clearing condition ((4.14) for $t = 1$) is divided by L_0

$$L_1 c_1^1 + L_0 c_1^2 = a_1 N_1 \left(k_1\right)^{\alpha} - K_2 \left| \frac{1}{L_0} \right. , \tag{4.34}$$

and using (3.2), (4.24), (3.3), and $G^{\tau} G^L \equiv G^n$ we get

$$G^L c_1^1 + c_1^2 = a_1 G^L \left(k_1\right)^{\alpha} - a_1 G^n G^L k_2. \tag{4.35}$$

Assume that the remaining variables within the system of intertemporal market equilibrium equations (4.27)–(4.35) can be substituted for the corresponding variables which are determined by the FOCs for short-run intergenerational efficiency. Under this proviso calculate $k_1 = (k_1)^{\circ}$ from (4.33), and you will get:

$$\left(k_1\right)^{\circ} = \frac{\left[(k_0)^{\circ}\right]^{\alpha}}{G^n} - \frac{(c_0^1)^{\circ}}{a_0 G^n} - \frac{(c_0^2)^{\circ}}{a_0 G^L G^n}, \tag{4.36}$$

which is, upon setting $G^L = G^{\tau} = G^n = 1 = a_{-1} = L_{-1}$, equal to (3.25). Next insert (4.36) into (4.29) and combine the result with (4.31) for $t = 1$ to get (3.29) for the case where $G^L, G^{\tau}, G^n, a_{-1} \neq 1$ and after using $1 + \theta \equiv 1/\beta$:

$$(1+\theta)\frac{\left(c_1^2\right)^\circ}{\left(c_0^1\right)^\circ} = \alpha\left\{G^{n^{-1}}\left[(k_0)^\circ\right]^\alpha - G^{n^{-1}}\frac{\left(c_0^1\right)^\circ}{a_0} - G^{n^{-1}}G^{L^{-1}}\frac{\left(c_0^2\right)^\circ}{a_0}\right\}^{\alpha-1}. \quad (4.37)$$

The final step is to insert the zero-period commodity market clearing condition (4.33) into the corresponding first-period condition (4.35) and substitute c_1^1 for $\underline{c}^1$ and k_2 for $\underline{k}$. Solving the resulting equation for $(c_1^2)^\circ$ we get the generalized version of (3.27) where $G^L, G^\tau, G^n, a_{-1} \neq 1$:

$$\left(c_1^2\right)^\circ = a_1 G^L\left\{\frac{\left[(k_0)^\circ\right]^\alpha}{G^n} - \frac{(c_0^1)^\circ}{a_0 G^n} - \frac{(c_0^2)^\circ}{a_0 G^n G^L}\right\}^\alpha - G^n \underline{c}^1 - a_1 G^n G^L \underline{k}. \quad (4.38)$$

□

Proposition 4.3, an application of the first theorem of welfare economics, is illustrated in Fig. 4.3. The figure shows the equivalence of the intertemporal market equilibrium allocation and short–run intergenerational efficiency.

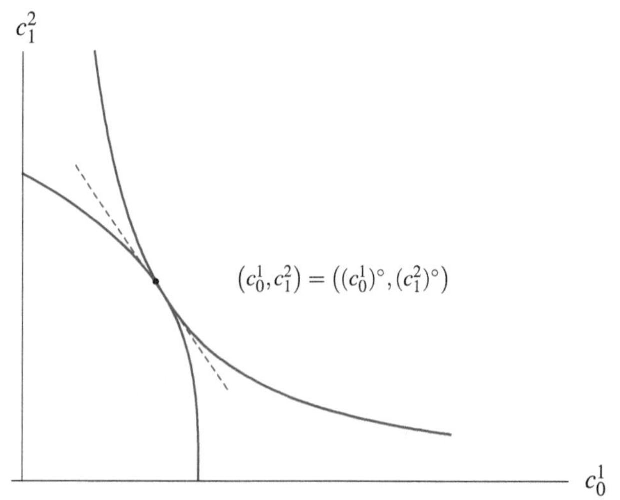

Fig. 4.3 Short–run intergenerational efficiency of the intertemporal market equilibrium

4.7 Conclusions

In this chapter we described extensively the institutional framework of perfectly competitive markets in our log-linear CD OLG economy with abundant natural resources. We derived the FOCs for intertemporal utility and for intratemporal profit maximization. Together with the market clearing conditions the intertemporal equilibrium dynamics of per–efficiency capital intensities and associated relative prices were derived. In the previous section we demonstrated an application of the first theorem of welfare economics to the intertemporal market equilibrium in the log-linear OLG economy. We showed that the conditions for an intertemporal market equilibrium implied those for short-run intergenerational efficiency.

The reader is reminded that in this chapter we were only able to demonstrate *short-run* intergenerational efficiency of the intertemporal market equilibrium. Thus, in the next chapter we will consider the *steady state* market equilibrium and *long-run* intergenerational efficiency.

References

Arrow K, Debreu G (1954) Existence of equilibrium for a competitive economy. Econometrica 22:265–290

Farmer K, Stadler I (2005) Marktdynamik und Umweltpolitik: Ein Beitrag zur gleichgewichts- und ordnungstheoretischen Fundierung umweltorientierter Volkswirtschaftslehre. LIT-Verlag, Wien

Hayek AF (1963) Arten der Ordnung. Ordo 14:1–30

Hayek FA (1973) Law, legislation, and liberty: Rules and order. Routledge & Kegan Paul, London

Hayek FA (1988) The fatal conceit: the errors of socialism. Routledge, London

Kirzner I (1973) Competition and entrepreneurship. University of Chicago Press, Chicago

Pareto V (1909) Manual d' économie politique pure. Girard & Briere, Paris

Ramsey F (1928) A mathematical theory of savings. Economic Journal 38:543–559

Walras L (1874) Elements of pure economics. Irwin, Homewood

Chapter 5
Steady-State Market Equilibrium, Long-Run Intergenerational Efficiency, and Optimality

5.1 Introduction

In the previous chapter we explored a competitive market economy with overlapping generations along an intertemporal equilibrium path. So far we have not been able to ascertain whether a market equilibrium with time–stationary per–efficiency capital intensities (i.e. a non–trivial steady state) does in fact exist, and if it exists, whether it is dynamically stable. These seminal questions will be investigated in the next section. Then we will define long-run intergenerational efficiency in our log-linear CD OLG economy, and provide a characterization of this efficiency notion. In the subsequent section the major question of this chapter will be explored, namely whether the steady state market equilibrium is long-run intergenerationally efficient. Moreover, the notion of intergenerational optimality will be introduced and compared to intergenerational efficiency. Finally, the role of resource augmenting technological progress for steady state economic growth is elaborated upon. As in Chaps. 3 and 4 we continue to assume here that both exhaustible and renewable natural resources are abundant and hence need not be considered in our model.

5.2 Steady-State Market Equilibrium

First, we have to define precisely what we mean by a *steady-state* market equilibrium. A steady-state market equilibrium exists when over an infinite horizon k_t approaches a finite, non-negative value k which can be calculated from the intertemporal equilibrium dynamics $k_{t+1} = (\sigma/G^n)(k_t)^\alpha$ by setting $k_{t+1} = k_t = k$.

Second, we need to investigate whether a non-negative, finite k exists and whether there is one or more steady-state solutions. The following proposition gives a precise answer.

K. Farmer and B. Bednar-Friedl, *Intertemporal Resource Economics: An Introduction to the Overlapping Generations Approach*, DOI 10.1007/978-3-642-13229-2_5,

Proposition 5.1 (Existence of steady states). *Within the log–linear CD OLG model with abundant natural resources, there exist exactly two steady state equilibria for* $0 < \alpha < 1$. *One solution comprises the trivial steady state, i.e.* $k = 0$, *and the other solution is given by the non–trivial steady state* $k = (\sigma/G^n)^{(1/(1-\alpha))}$.

Proof. The fixed points of the fundamental equation of motion of the intertemporal equilibrium dynamics (4.24), i.e. where $k_{t+1} = k_t = k$, result from the solution of the following equation

$$G^n k = \sigma k^\alpha, \tag{5.1}$$

and they are as follows: $k = 0$ and $k = \left(\frac{\sigma}{G^n}\right)^{\frac{1}{1-\alpha}}$. □

From the existence proposition follows an important corollary which better clarifies the notion 'steady state'. The corollary concerns the GDP growth rate associated with the capital intensity at the non-trivial fixed point. GDP growth is steady-state since the growth rate does not change over time.

Corollary 5.1. *The steady-state GDP growth rate is determined as follows:* $g^Y = G^n - 1$. *Moreover, the growth rate of the per-capita product* $g^{Y/L} = G^\tau - 1$ *is independent of the population growth rate.*

Proof.

$$g_t^Y = G_t^Y - 1 = \frac{Y_{t+1}}{Y_t} - 1 = \frac{(A_{t+1})^{1-\alpha}(K_{t+1})^\alpha}{(A_t)^{1-\alpha}(K_t)^\alpha} - 1 == G^n \left(\frac{k_{t+1}}{k_t}\right)^\alpha - 1 \tag{5.2}$$

For $k_{t+1} = k_t = k$, the GDP growth rate is thus $g^Y = G^n - 1$, and the growth rate of the per-capita product is

$$g_t^{Y/L} = \frac{G_t^Y}{G^L} - 1 = \frac{G^L G^\tau}{G^L} - 1 = G^\tau - 1. \tag{5.3}$$

□

The next question concerns the *dynamic stability* of both steady state solutions. The answer is provided by the following proposition. Before we state this proposition we have to clarify the term 'asymptotic stability'. Loosely speaking, it means that the capital intensity dynamics starting from a capital intensity lower (or larger) than the steady state, automatically approaches the steady state as time goes to infinity. Otherwise, the dynamics is asymptotically unstable. As we will see asymptotic stability depends on the magnitude of the differential quotient of period $t+1$ capital intensity to that in period t, i.e. $\mathrm{d}k_{t+1}/\mathrm{d}k_t$. If this quotient lies between zero and one, capital intensity dynamics is asymptotically stable, otherwise it is asymptotically unstable.

Proposition 5.2 (Dynamic stability of steady states). *The trivial steady state is asymptotically unstable, while the non-trivial steady state is asymptotically stable.*

Proof. To prove this proposition we first calculate $\mathrm{d}k_{t+1}/\mathrm{d}k_t$ from (4.24). It is easily seen that $\mathrm{d}k_{t+1}/\mathrm{d}k_t = \alpha\sigma/G^n k_t^{\alpha-1}$. Clearly, $\mathrm{d}k_{t+1}/\mathrm{d}k_t$ is larger than one (and the dynamics is unstable) if $k_t = 0$. On the other hand if $k_t > 0$ the differential quotient is larger than zero. It is also less than one at least in a small neighborhood of $k = \left(\frac{\sigma}{G^n}\right)^{\frac{1}{1-\alpha}}$, since for $k_t = k$ we have $\mathrm{d}k_{t+1}/\mathrm{d}k_t = \alpha$, which is less than one by assumption. □

Figure 5.1 illustrates the stability of the non-trivial steady-state solution k. For initial capital stocks smaller than k, the capital stock increases towards its steady-state value while it decreases for initial capital stocks larger than k.

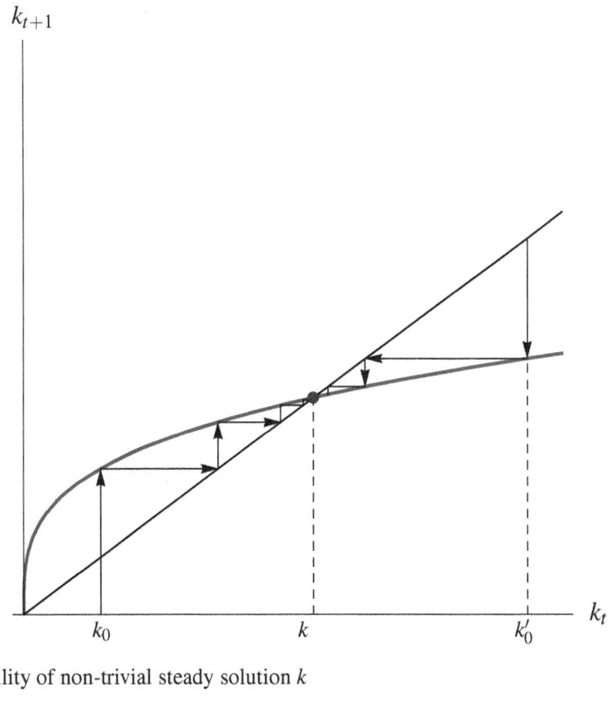

Fig. 5.1 Stability of non-trivial steady solution k

In a competitive market system capital intensities are determined by firms in response to relative factor prices. It is interesting to examine how firms are led by market signals to adapt non-steady state capital intensities such that the economy approaches the non-trivial steady state without the directives of a central planner. Thus, assume the initial capital intensity k_0 is less than non-trivial steady-state capital intensity k. Since $g_0^K = K_1/K_0 - 1 = (K_1/A_1)/[(A_0/A_1)/(K_0/A_0)] - 1 = k_1/(k_0/G^n) - 1 = \sigma(1/k_0)^{(1-\alpha)} - 1$ is larger than the corresponding steady-state growth factor of capital, given by $g^K = \sigma(1/k)^{(1-\alpha)} - 1 = g^A = g^n$ because $k_0 < k$.

In other words, in period zero capital grows faster than efficiency labor, and labor becomes scarce relative to capital. As a consequence, real wage increases relative to real capital price. As we know from the FOCs for profit maximization, profit maximizing firms respond to a rising relative wage rate by increasing capital intensity, and this is the right decision with regard to period-one capital intensity, when the initial capital intensity is too low relative to steady-state capital intensity (see also Farmer and Wendner, 1999, 78).

5.3 Long-Run Intergenerational Efficiency

This section is devoted to clarifying the reason behind long-run intergenerational inefficiency of market allocations in economies with overlapping generations. As de la Croix and Michel (2002) point out, when there is an infinity of goods and agents, two aspects of long-run intergenerational efficiency need to be distinguished: (i) dynamic (in-)efficiency of production when the production frontier is extended to an infinite horizon set up; and (ii) whether the type of generation considered in steady state is the younger generation only or also the initial older generation.

Let us start with the problem of dynamic efficiency of production in the infinite horizon set up. As in Chap. 3 we first define a feasible steady-state capital intensity, before deriving the related efficiency conditions.

Definition 5.1 (Steady state feasibility). The per-efficiency capital capital stock $k \geq 0$ is feasible in the steady state if per-efficiency capita production k^{α} is at least large enough to allow for replacement investment $G^n k$.

It is convenient here to define the difference between production and replacement investment as 'net production', i.e. $\phi(k) \equiv k^{\alpha} - G^n k$. Moreover, let $\bar{k}$ be the per–efficiency capital intensity k where net production is zero. It is easy to see that $\bar{k} = (G^n)^{1/(\alpha-1)}$. This enables us to identify that level of capital intensity which maximizes available consumption for the generations living in steady state.

Proposition 5.3 (Golden rule). *For all $k \in (0, \bar{k})$ per–efficiency capital is steady-state feasible. Moreover, there exists a unique capital intensity k^{GR} such that $\alpha k^{GR^{\alpha-1}} = G^n$, which is called the 'Golden rule' capital intensity.*

Having identified the Golden rule capital intensity, we are now in a position to state the prerequisites for steady-state dynamic efficiency.

Proposition 5.4 (Steady-state dynamic efficiency). *A capital intensity $k \in (0, \bar{k})$ is steady-state dynamically efficient if $k < k^{GR}$, while it is steady-state dynamically inefficient if $k > k^{GR}$. If $\alpha \geq 1/2$, the capital intensity is steady-state dynamically efficient.*

Proof. According to lemma 1 of Galor and Ryder (1991, 388), capital intensity k is steady-state dynamically efficient if $k \geq [(1-\alpha)/G^n]^{(1/(1-\alpha))}$ for $k \geq k^{GR}$. This is the case if $\alpha \geq (1/2)$. □

Thus, without assuming $\alpha \geq 1/2$ we cannot prove intergenerational efficiency of the consumption allocation in an infinitely lasting intertemporal market equilibrium. Figure 5.2 depicts a case in which there exist capital intensities for which $k < k^{GR} < \bar{k}$, and hence these capital intensities are dynamically efficient.

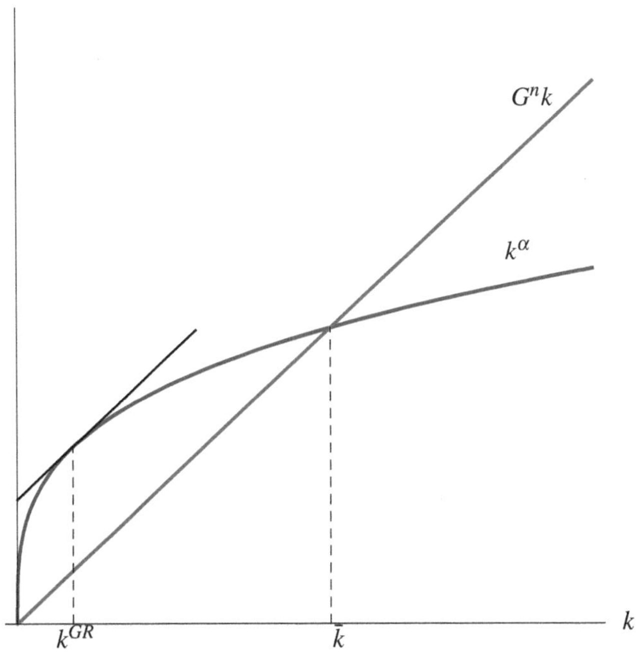

Fig. 5.2 Dynamically efficient capital intensities $k < k^{GR}$ and Golden rule capital intensity $k^{GR} < \bar{k}$

Having investigated the steady-state dynamic efficiency of production, let us move on to examine long-run intergenerational efficiency including the household side. Here we encounter now the second problem mentioned above, namely whether the initial old generation is included in the efficiency calculus or not. In his seminal article, Diamond (1965, 1128-1129) defines 'golden age' paths by excluding the initial older generation. Here we include the initial older generation since this follows naturally from the first definition of intergenerational efficiency provided in Chap. 3, where the utility of the initial older generation is maximized subject to the constraint that the subsequent younger generations obtain certain utility levels and that the aggregate consumption and investment constraint is met.

However, in this chapter we assume time-stationary values for capital intensities and for the consumption variables of all younger generations. Hence, the infinite number of younger generations' utility constraints collapses to just one. As is well-known from standard microeconomics it does not matter for the FOCs for Pareto efficiency whose household's utility function is maximized (Mas-Colell et al, 1995,

562). Hence, in order to obtain Diamond's 1965 'golden age' paths as special cases, we interchange the role of the older and the younger household within the efficiency calculus: the life-cycle utility of the younger generations is maximized subject to the constraints that the initial older generation obtains a certain utility level and that the aggregate consumption investment constraint is met. For ease of exposition, we restate the steady-state consumption levels in per-efficiency capita notation, i.e. $\tilde{c}^1 \equiv c^1/a_\infty$, $\tilde{c}^2 \equiv c^2/a_\infty$, and $\tilde{c}_0^2 \equiv c_0^2/a_\infty$. We start by defining long-run intergenerational efficiency.

Definition 5.2 (Long-run intergenerational efficiency). The capital intensity $k^\bullet$, the intertemporal consumption bundle of the younger generation in the steady state $\{(\tilde{c}^1)^\bullet, (\tilde{c}^2)^\bullet\}$, and the consumption $(\tilde{c}_0^2)^\bullet$ of the initial older generation are long-run intergenerationally efficient if $(k^\bullet, (\tilde{c}^1)^\bullet, (\tilde{c}^2)^\bullet, (\tilde{c}_0^2)^\bullet)$ is the solution of the following maximization calculus:

$$\begin{aligned}
&\max \quad \ln \tilde{c}^1 + \beta \ln \tilde{c}^2 \\
&\text{subject to:} \\
&\ln \tilde{c}_0^2 \geq \ln(\tilde{c}_0^2)^\bullet, \\
&\tilde{c}^1 + \tilde{c}_0^2/(G^L) \leq (k_0)^\alpha - G^n k, \\
&\tilde{c}^1 + \tilde{c}^2/G^L \leq k^\alpha - G^n k.
\end{aligned}$$

The corresponding Lagrangian to the above stated problem reads as:

$$\begin{aligned}
\mathscr{L} = \ln \tilde{c}^1 + \beta \ln \tilde{c}^2 + \mu_{-1}^c[\ln \tilde{c}_0^2 - \ln(\tilde{c}_0^2)^\bullet] + \phi_0^y[(k_0)^\alpha - G^n k - \tilde{c}^1 - \tilde{c}_0^2/G^L] + \\
+ \phi^y[k^\alpha - G^n k - \tilde{c}^1 - \tilde{c}^2/G^L], \quad (5.4)
\end{aligned}$$

where μ_{-1}^c, ϕ_0^y, and ϕ^y denote the Lagrangian multipliers. Upon differentiating this function, we obtain the FOCs for long-run intergenerational efficiency as summarized in the following proposition.

Proposition 5.5 (Long-run intergenerational efficiency). *Long-run intergenerationally efficient capital intensity $k^\bullet$, the intertemporal consumption bundle of younger generation $\{(\tilde{c}^1)^\bullet, (\tilde{c}^2)^\bullet\}$ and the consumption $(\tilde{c}_0^2)^\bullet$ of the initial older generation can be determined by solving the following FOCs:*

$$(\mu_{-1}^c)^\bullet/(\tilde{c}_0^2)^\bullet = (\phi_0^y)^\bullet/G^L, \tag{5.5}$$

$$1/(\tilde{c}^1)^\bullet = (\phi_0^y)^\bullet + (\phi^y)^\bullet, \tag{5.6}$$

$$\beta/(\tilde{c}^2)^\bullet = (\phi^y)^\bullet/(G^L), \tag{5.7}$$

$$\alpha(k^\bullet)^{\alpha-1} = G^n(1 + (\phi_0^y)^\bullet/(\phi^y)^\bullet), \tag{5.8}$$

$$(\tilde{c}^1)^\bullet + (\tilde{c}_0^2)^\bullet/G^L = (k_0)^\alpha - G^n k^\bullet, \tag{5.9}$$

$$(\tilde{c}^1)^\bullet + (\tilde{c}^2)^\bullet/G^L = (k^\bullet)^\alpha - G^n k^\bullet. \tag{5.10}$$

Proof. Differentiating (5.4) with respect to all endogenous variables and with respect to the Lagrangian multipliers, and setting the partial derivatives equal to zero gives (5.5)–(5.10). □

Diamond's 'golden age' results as a prominent special case of the FOCs for long-run intergenerational efficiency if we assume that μ_{-1}^{c} is equal to zero, or in other words, that the utility demands of the initial older generation and the transition from a historically given capital intensity towards the efficient intensity are ignored. This leads us to the following proposition stating the FOCs for golden age paths of long-run intergenerational efficiency.

Corollary 5.2 (Golden-age paths for long-run intergenerational efficiency). *Suppose that* $(\mu_{-1}^{c})^{\bullet} = 0$. *Then, the golden age capital intensity is equal to the golden rule capital intensity following from* $\alpha(k^{GR})^{\alpha-1} = G^n$. *The golden-age intertemporal consumption bundle of the younger household* $\{(\tilde{c}^1)^{\bullet}, (\tilde{c}^2)^{\bullet}\}$ *is determined by the following FOCs:*

$$(\tilde{c}^2)^{\bullet}/(\beta(\tilde{c}^1)^{\bullet}) = G^L, \tag{5.11}$$

$$\phi(k^{GR}) = (\tilde{c}^1)^{\bullet} + (\tilde{c}^2)^{\bullet}/G^L. \tag{5.12}$$

Having thus characterized long-run intergenerational efficiency we are now ready to ask whether the steady-state market equilibrium capital intensity k, and the intertemporal and intergenerational consumption allocation, are long-run intergenerationally efficient. The next section is devoted to answering this question.

5.4 Long-Run Intergenerational (In-)Efficiency of Steady-State Market Equilibrium

In the previous chapter we have shown that the intertemporal market equilibrium in a finite-horizon economy with overlapping generations is short-run intergenerationally efficient. It would thus appear natural to suggest that this is also true in the long run. However, Diamond (1965) already demonstrated that a steady-state market equilibrium might not be long-run intergenerationally efficient.

To confirm Diamond's assertion, we first reiterate the relevant steady-state conditions of the competitive market economy by again using per efficiency capita notation:

$$\tilde{c}_0^2 = G^L \alpha (k_0)^{\alpha}, \tag{5.13}$$

$$\tilde{c}^1 + \tilde{c}^2/r = w, \tag{5.14}$$

$$\tilde{c}^2/(\beta \tilde{c}^1) = r, \tag{5.15}$$

$$(1-\alpha)k = w, \tag{5.16}$$

$$\alpha k^{\alpha-1} = r, \tag{5.17}$$

$$k^{\alpha} = \tilde{c}^1 + \tilde{c}^2/G^L + G^n k \iff G^n k = \sigma k^{\alpha}. \tag{5.18}$$

In order to be able to check whether the steady-state market equilibrium is long-run intergenerationally efficient, we set $(\tilde{c}_0^2)^{\bullet} = \tilde{c}_0^2 = G^L\alpha(k_0)^{\alpha}$. Imposing this equality is feasible since it still leaves $(\tilde{c}_0^2)^{\bullet}$ undetermined within the efficiency calculus.

The next step is to reduce both the market equilibrium (5.13)–(5.18) and the efficiency conditions (5.5)–(5.10) to three equations in order to determine, on the one hand, the three endogenous market variables, k, $\tilde{c}^1$, $\tilde{c}^2$, and on the other, the three long-run intergenerational efficiency variables, $k^{\bullet}$, $(\tilde{c}^1)^{\bullet}$, $(\tilde{c}^2)^{\bullet}$.

From the steady-state market equilibrium conditions it is easy to see that the three endogenous market equilibrium variables are determined as follows:

$$k = (\sigma/G^n)k^{1/(1-\alpha)}, \quad \tilde{c}^1 = (1-\alpha)/(1+\beta)k^{\alpha}, \quad \tilde{c}^2 = \alpha k^{\alpha-1}\beta\tilde{c}^1.$$

While it is not so easy to determine the three efficiency variables a little manipulation of the efficiency conditions reveals the following three (implicit) equations:

$$(1-\alpha)(k_0)^{\alpha}(1+\alpha\beta(k^{\bullet})^{\alpha-1}/G^L) = (k^{\bullet})^{\alpha}(1+\alpha\beta), \tag{5.19}$$

$$(\tilde{c}^1)^{\bullet} = (1-\alpha)(k_0)^{\alpha} - G^n k^{\bullet}, \tag{5.20}$$

$$(\tilde{c}^2)^{\bullet} = \alpha(k^{\bullet})^{\alpha-1}\beta(\tilde{c}^1)^{\bullet}. \tag{5.21}$$

At first sight it is not apparent whether the steady-state market solution is long-run intergenerationally efficient or not. However, if we assume that the initial capital intensity is occasionally equal to the long-run intergenerationally efficient capital intensity, i.e. $k_0 = k^{\bullet}$, then the first of the three efficiency equations above implies that $k^{\bullet} = k$. This result is not surprising since we assumed that the market economy started already at the long-run intergenerationally efficient capital intensity. It is also immediately clear that for an initial capital intensity differing from the long-run efficient capital intensity the market solution turns out to be long-run intergenerationally inefficient. Indeed, the reader can check by the use of numerically specified parameters that $k_0 < k^{\bullet} \Rightarrow k > k^{\bullet}$ and $k_0 > k^{\bullet} \Rightarrow k < k^{\bullet}$, i.e. that when the initial capital intensity is less than the long-run efficient capital intensity the competitive market economy 'over-accumulates' capital, and otherwise it 'under-accumulates' capital.

For the special case that both the initial capital intensity and the utility level of the initial older generation are ignored in calculating the long-run intergenerationally efficient solution, we are now ready to verify the Diamond (1965) claim that a competitive market economy does not end up in an intergenerationally efficient allocation.

The FOCs for the golden age paths above now represent the benchmark for the market solution. We know that

$$k^{\bullet} = k^{GR} = (\alpha/G^n)^{1/(1-\alpha)}, \tag{5.22}$$

$$(\tilde{c}^2)^{\bullet}/(\beta(\tilde{c}^1)^{\bullet}) = G^L, \tag{5.23}$$

$$(k^{GR})^{\alpha} - G^n k^{GR} = (\tilde{c}^1)^{\bullet} + (\tilde{c}^2)^{\bullet}/G^L. \tag{5.24}$$

Since $(k^{GR})^\alpha - G^n k^{GR} = (1-\alpha)(k^{GR})^\alpha$,

$$(\tilde{c}^1)^\bullet = (\tilde{c}^1)^{GR} = (1/(1+\beta))(1-\alpha)(k^{GR})^\alpha,$$

and

$$(\tilde{c}^2)^\bullet = (\tilde{c}^2)^{GR} = (\beta/(1+\beta))G^L(1-\alpha)(k^{GR})^\alpha,$$

the corresponding consumption quantities of the market solution shown above will coincide with the efficient quantities if $k^{GR} = k$ or equivalently if $(\alpha/G^L)^{1/(1-\alpha)} = (\sigma/G^L)^{1/(1-\alpha)}$. Clearly, this is only true if $\sigma = \alpha$ or $\beta/(1+\beta) = \alpha/(1-\alpha)$.

To demonstrate the long-run intergenerational efficiency of Golden rule capital intensity via an alternative route, start with the asset market equilibrium and the intertemporal budget constraint and then insert the profit maximizing real wage rate and real capital price into the equations. This gives:

$$G^n k = (1-\alpha)k^\alpha - \tilde{c}^1, \tag{5.25}$$

$$\tilde{c}^1 + \tilde{c}^2/(\alpha k^{\alpha-1}) = (1-\alpha)k^\alpha. \tag{5.26}$$

Solving the first of these two equations for $\tilde{c}^1$ and inserting the result into the second equation, gives k as a positively sloped function of $\tilde{c}^2$, denoted by $k = k(\tilde{c}^2)$:

$$k \equiv k(\tilde{c}^2) = (1/(\alpha G^n))^{1/\alpha}(\tilde{c}^2)^{1/\alpha}, \quad \text{with} \quad \mathrm{d}k/\mathrm{d}\tilde{c}^2 = 1/(\alpha^2 G^n k^{\alpha-1}) > 0. \tag{5.27}$$

Inserting this solution for k back into the first of the equations, we get a relation between $\tilde{c}^2$ and $\tilde{c}^1$ which is termed the 'consumption possibility frontier' in a $(\tilde{c}^2, \tilde{c}^1)$–diagram:

$$\tilde{c}^1 = (1-\alpha)(1/(\alpha G^n)\tilde{c}^2 - G^n\left(1/(\alpha G^n)\right)^{1/\alpha}\left(\tilde{c}^2\right)^{1/\alpha}. \tag{5.28}$$

This frontier is depicted in Fig. 5.3. The (negative) slope of this curve is $-\mathrm{d}\tilde{c}^2/\mathrm{d}\tilde{c}^1 = (\alpha^2 G^n k^{a-1})/(G^n + \alpha(\alpha-1)k^{\alpha-1})$. Clearly, if the capital intensity is Golden rule, that is $\alpha k^{\alpha-1} = G^n$, the (negative) slope of the consumption possibility frontier equals G^n.

The curve confronting the consumption possibility frontier is the intertemporal indifference curve which is also depicted in Fig. 5.2. Its slope is the marginal rate of substitution $-(\mathrm{d}\tilde{c}^2/\mathrm{d}\tilde{c}^1) = \tilde{c}^2/(\beta\tilde{c}^1)$. In the golden rule intertemporal utility maximum, $-(\mathrm{d}\tilde{c}^2/\mathrm{d}\tilde{c}^1) = \tilde{c}^2/(\beta\tilde{c}^1) = r = \alpha k^{\alpha-1} = G^n$. Thus, the intertemporal indifference curve most distant from the origin is tangential to the consumption possibility frontier at the golden rule capital intensity.

However, there is no reason why in general the aggregate saving rate should be equal to the production elasticity of capital, in which case the market economy need not end up with an intergenerationally efficient consumption allocation. In particular, if the saving rate is larger than the production elasticity of capital, there is 'over-saving' and 'over-accumulation' of capital, leading to a dynamically inefficient level of capital intensity in the market economy. If, on the other hand, $\sigma < \alpha$ there is 'under-saving' and 'under-accumulation' of capital in the market economy,

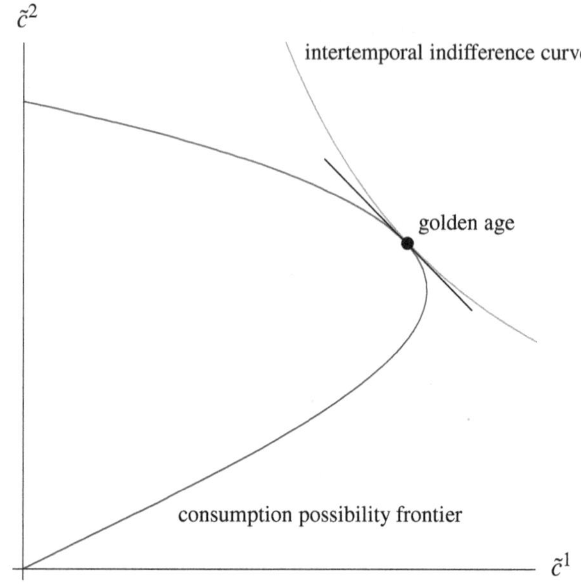

Fig. 5.3 Intergenerational efficiency and the Golden rule

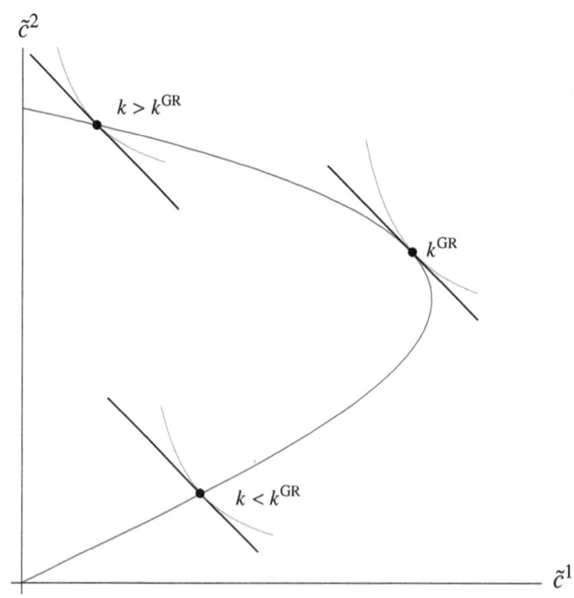

Fig. 5.4 Intergenerational (in)efficiency of steady-state market equilibria

resulting in a dynamically efficient level of the capital intensity in the competitive economy. Both of these cases are depicted in Fig. 5.4.

Note, however, that while dynamic efficiency of the capital intensity is a necessary condition for long-run intergenerational efficiency, it is not sufficient. On the other hand, dynamic inefficient capital intensity is incompatible with long-run intergenerational efficiency since under such circumstances the welfare of at least one generation can in principle be increased without decreasing the utility of any other generation.

5.5 Intergenerational Efficiency versus Intergenerational Optimality

Intergenerational efficiency does not require that the utilities of subsequent generations be directly comparable. While this is clearly a methodological advantage, it has a drawback in that nothing can then be said about the overall welfare balance between generations. Furthermore, even if we accept that the utilities of different generations need not be summed up for purposes of comparison, the question of how or whether alternative institutional frameworks and policy interventions increase or decrease overall welfare remains extremely important. Unless we answer this question, how else are we to characterize an optimal policy? Thus, in the present analysis, in order to better understand the difference between intergenerational optimality and intergenerational efficiency, we first derive the FOCs for intergenerational optimality and then compare them to those for intergenerational efficiency.

To derive the conditions for intergenerational optimality, we assume the existence of a benevolent social planner whose objective is to maximize a discounted sum of life cycle utility for all current and future generations. This gives:

$$\max \quad \sum_{t=-1}^{\infty} \gamma^t (\ln c_t^1 + \beta \ln c_{t+1}^2)$$

$$\text{subject to}$$

$$(k_t)^\alpha = G^n k_{t+1} + c_t^1 + c_t^2 / G^L,$$

whereby k_0 and c_{-1}^1 are exogenously (historically) given and γ is the social discount factor, assumed here to be less than or equal to unity. While for $\gamma = 1$ the social planner's objective function is not defined, there are formal ways to circumvent this problem (see de la Croix and Michel, 2002, 92). However, for the sake of simplicity we will focus on the case of $\gamma < 1$.

Since the life-cycle utility function is additively separable, we can rearrange the objective function of the planner in the following way:

$$\sum_{t=0}^{\infty} \gamma^t \left[\ln c_t^1 + \beta/\gamma \ln c_t^2\right]. \tag{5.29}$$

One simple method to obtain the FOCs for intergenerational optimality is to solve the aggregate consumption and investment constraint from above for c_{t+1}^2 and then to insert the solution into the rearranged objective function of the planner. This leads to the following unconstrained maximization problem:

$$\sum_{t=0}^{\infty} \gamma^t (\ln c_t^1 + \beta/\gamma \ln(G^L[(k_t)^\alpha - G^n k_{t+1} - c_t^1])). \tag{5.30}$$

The FOCs are obtained by differentiating this objective function with respect to c_t^1 and k_{t+1}:

$$1/c_t^1 = \beta\gamma^{-1}(G^L/c_t^2), \tag{5.31}$$

$$G^n/c_t^2 = \alpha\gamma(k_{t+1})^{\alpha-1}/c_{t+1}^2. \tag{5.32}$$

Together with the aggregate consumption and investment constraint, $G^n k_{t+1} = (k_t)^\alpha - c_t^1 - c_t^2/G^L$, the intergenerationally optimal paths for k_{t+1}, c_t^1, c_t^2 can be found by means of the method of unknown coefficients. Suppose that $c_t^1 = \xi_1(k_t)^\alpha$ and $c_t^2 = \xi_2(k_t)^\alpha$ whereby ξ_1 and ξ_2 are as yet unknown coefficients. Inserting these equations into (5.32), gives: $G^n k_{t+1} = \alpha\gamma(k_t)^\alpha$. Inserting this result into the aggregate consumption and investment constraint together with the hypotheses for c_t^1 and c_t^2, we obtain $\alpha\gamma = 1 - \xi_1 - \xi_2/G^L$. From the insertion of the hypotheses for c_t^1 and c_t^2 into (5.31), we obtain $\xi_2 = \beta\gamma^{-1}G^L\xi_1$. Combining this equation with $\alpha\gamma = 1 - \xi_1 - \xi_2/G^L$, we get $\xi_1 = (1-\alpha\gamma)/(1+\beta\gamma^{-1})$ and $\xi_2 = \beta\gamma^{-1}G^L\xi_1$.

Hence, the intergenerationally optimal path (denoted by $\star$) for the capital intensity and younger and older consumption reads as follows:

$$k_{t+1}^\star = (\alpha\gamma)/G^n (k_t^\star)^\alpha, \tag{5.33}$$

$$(c_t^1)^\star = (\frac{(1-\alpha\gamma)}{(1+\beta\gamma^{-1})}(k_t^\star)^\alpha, \tag{5.34}$$

$$(c_t^2)^\star = \beta\gamma^{-1}\frac{(1-\alpha\gamma)}{(1+\beta\gamma^{-1})}G^L(k_t^\star)^\alpha. \tag{5.35}$$

This optimal solution also fulfills the so-called 'transversality condition'

$$\lim_{t\to\infty} \gamma^t [(c_t^1)^\star]^{-1} k_{t+1}^\star = 0,$$

which requires that the limit of the marginal utility of capital in terms of consumption be nil (see de la Croix and Michel, 2002, 103).

On looking now for a steady-state intergenerationally optimal solution to the planner's problem, denoted by $k^\star, (c^1)^\star, (c^2)^\star$, we get immediately from the optimal dynamics above:

$$k^\star = ((\alpha\gamma)/G^n)^{1/(1-\alpha)} \Leftrightarrow \alpha(k^\star)^{\alpha-1} = G^n/\gamma, \tag{5.36}$$

$$(c^1)^\star = \frac{(1-\alpha\gamma)}{(1+\beta\gamma^{-1})}(k^\star)^\alpha, \tag{5.37}$$

$$(c^2)^\star = \beta\gamma^{-1}\frac{(1-\alpha\gamma)}{(1+\beta\gamma^{-1})}G^L(k^\star)^\alpha. \tag{5.38}$$

The first equation states the so-called ‘modified Golden rule’. This says that marginal productivity of steady-state capital intensity equals the natural growth factor divided by the planner’s discount factor. Clearly, if the discount factor is equal to one the intergenerational efficient capital intensity coincides with the intergenerational optimal capital intensity.

We thus arrive at the optimality condition for intergenerational allocation (see (5.31)) namely: the marginal utility of the younger generation is equal to the marginal utility of the older generation times the population growth factor multiplied by the individual discount factor divided by the social discount factor, or, in other words, the intertemporal marginal rate of substitution between individual present and future consumption is equal to the planner’s discount factor divided by the population growth factor.

So far in this chapter we have assumed that natural resources are abundant. This is clearly a digression from our main topic. Thus, to return to the role of the significance of natural resources for economic growth we close this chapter by focusing again on scarce exhaustible resources and by investigating the role of resource augmenting technological progress for steady economic growth with population growth.

5.6 Steady-State Economic Growth and Resource Saving Technological Progress

If natural resources (exhaustible and renewable) are free (not scarce) then the steady-state growth rate of the gross domestic product is exclusively determined by (it is equal to) the sum of the population growth rate and the rate of labor augmenting technological progress (labor productivity). But what are the determinants of long-run economic growth if natural resources (in particular exhaustible resources) secure a scarcity rent which enters economic costs and hence influences the market prices of produced commodities?

Taking account of the dependence of commodity production on scarce natural resources, does not seem to help here since it tends to stifle the role played by labor (capital) saving technological progress as a substitute for natural resources in economic growth. To provide a simple but insightful answer to the question above we therefore focus on neutral technological progress as defined by Hicks (1932). Neutral technological progress in the sense of Hicks means that the marginal products of all factors change by the same amount assuming the factor intensities do not change.

In order to keep the analysis of such technological progress as simple as possible, we again presuppose the use of an aggregate CD production function with Hicks–neutral technological progress, with respect to not only labor N, and man-made capital K, but also with respect to natural (exhaustible) resources S. Thus:

$$Y(t) = e^{g^{\tau}t} K(t)^{\alpha_3} N(t)^{\alpha_2} S(t)^{\alpha_1}, \quad \alpha_3 + \alpha_2 + \alpha_1 = 1. \tag{5.39}$$

Here α_3, α_2 and α_1 denote the production elasticities of man-made capital, labor and the natural resource respectively. As usual, they are assumed to lie between zero and one and to sum to unity (constant returns to scale). Several things need to be noted with respect to the specification of this aggregate production function with Hicks-neutral technological progress.

Notice first that the production period for which this CD function is specified is infinitesimally small and hence the output and the inputs in the production process can be specified at any point in time, with time being specified as an explicit variable. As a consequence, all variables can be specified as continuously differentiable functions of time t indicated by $x(t)$ with $x = Y, K, N, S$. Second, in contrast to the discrete-time specification of technological progress in Chap. 3 above, we now assume a continuous-time flow of exogenous technological improvements depicted by $e^{g^\tau t}$, the limiting value of $(1+g^\tau/\nu)^{t\nu}$ when the number of subperiods ν goes to infinity. Third, the fact that $e^{g^\tau t}$ applies to all production factors in the production function implies equi-proportional technological progress with respect to all factors.

To move on from output and input levels to the growth rates of the produced commodity and the factor inputs, we next take the natural logarithms with respect to all variables in the aggregate production function:

$$\ln Y(t) = g^\tau t + \alpha_3 \ln K(t) + \alpha_2 \ln N(t) + \alpha_1 \ln S(t). \tag{5.40}$$

Differentiating this equation on both sides with respect to time, we get:

$$\frac{1}{Y(t)}\frac{\mathrm{d}Y(t)}{\mathrm{d}t} = g^\tau + \alpha_3 \frac{1}{K(t)}\frac{\mathrm{d}K(t)}{\mathrm{d}t} + \alpha_2 \frac{1}{N(t)}\frac{\mathrm{d}N(t)}{\mathrm{d}t} + \alpha_1 \frac{1}{S(t)}\frac{\mathrm{d}S(t)}{\mathrm{d}t}. \tag{5.41}$$

Setting $(1/x(t))\mathrm{d}x(t)/\mathrm{d}t = g^x(t)$, this equation can equivalently be written as follows:

$$g^Y(t) = g^\tau + \alpha_3 g^K(t) + \alpha_2 g^N(t) + \alpha_1 g^S(t). \tag{5.42}$$

We now assume steady-state growth, which implies that in the growth equation, all time references can be deleted:

$$g^Y = g^\tau + \alpha_3 g^K + \alpha_2 g^N + \alpha_1 g^S. \tag{5.43}$$

We know that in the steady-state the following growth relation holds: $g^K = g^Y$. Moreover, in the long run labor markets are cleared, i.e. $g^N = g^L$ and non-renewable resources are either exhausted or are conserved which implies that $g^S = 0$. Acknowledging these insights in equation (5.43), we obtain:

$$g^Y(1-\alpha_3) = g^\tau + \alpha_2 g^L, \tag{5.44}$$

and on considering $1-\alpha_3 = \alpha_2 + \alpha_1$, the expression for GDP growth rate is found to be:

$$g^Y = \frac{g^\tau}{\alpha_2 + \alpha_1} + \frac{\alpha_2}{\alpha_2 + \alpha_1} g^L. \tag{5.45}$$

From this, the growth of GDP per capita follows as:

$$g^{Y/L} = g^Y - g^L = \frac{g^\tau}{\alpha_2 + \alpha_1} - \frac{\alpha_1}{\alpha_2 + \alpha_1} g^L. \tag{5.46}$$

This equation shows that the growth rate of per capita income depends negatively on the population growth rate—an insight closely related to the growth pessimism exhibited by most classical economists, in particular by Malthus. In contrast to this, if natural resources do not have any productivity effects, i.e. $\alpha_1 = 0$, then the growth rate of per capita income remains independent of population growth, as found in neoclassical growth theory.

(5.46) also shows that the following inequality holds

$$g^{Y/L} > 0 \Leftrightarrow g^L < \frac{g^\tau}{\alpha_1}. \tag{5.47}$$

This equivalence says that the growth rate of income (output) per capita is larger than zero if and only if the population growth rate is less than the rate of resource saving technological progress divided by the production elasticity of the natural resource (Schmitt-Rink, 1990; Arnold, 1993). A lower population growth rate, a higher growth rate of resource saving technological growth, or a lower production elasticity of the natural resource (this indicates the importance of the natural resource in the production process) all increase the likelihood of a positive per capita growth rate.

5.7 Conclusions

This chapter is devoted to the investigation of the prospects for our log-linear OLG economy in the long run. First, we proved that the mild restriction of $0 < \alpha < 1$ implied the existence of a non-trivial steady state solution to the intertemporal equilibrium dynamics. This steady state is moreover asymptotically stable. Second, in contrast to the main result of the previous chapter, the steady-state market equilibrium is in general not long-run intergenerationally efficient. Here we encounter a major difference to the ILA approach. Third, as in classical growth theory without technological progress, in a neoclassical growth model with productive natural resources and resource saving technological progress population growth endangers the rise of living standards.

Clearly, neoclassical growth theory becomes more realistic once natural resources are inserted into the aggregate production function. Nevertheless, we still have to deal with the questions of intertemporal optimality and intergenerational efficiency with respect to renewable resources in a growing economy. The next chapters are devoted to elaborating on these important topics in intertemporal resource economics.

References

Arnold M (1993) Wachstum und ökologisches Gleichgewicht. No. 14 in Dynamische Wirtschaftstheorie, Lang, Frankfurt am Main

de la Croix D, Michel P (2002) A Theory of Economic Growth. Dynamics and Policy in Overlapping Generations. Cambridge University Press, Cambridge, New York

Diamond P (1965) National debt in a neoclassical growth model. American Economic Review 55:1126–1150

Farmer K, Wendner R (1999) Wachstum und Aussenhandel: Eine Einführung in die Gleichgewichtstheorie der Wachstums- und Außenhandelsdynamik. Springer, Heidelberg

Galor O, Ryder HR (1991) Dynamic efficiency of steady-state equilibria in an overlapping generations model with productive capital. Economic Letters 35:385–390

Hicks JR (1932) The theory of wages. Macmillan, London

Mas-Colell A, Whinston MD, Green JR (1995) Microeconomic Theory. Oxford University Press, Oxford

Schmitt-Rink G (1990) Begrenzte natürliche Ressourcen in der Wachstums- und Verteilungstheorie. WISU 6:377–392

Part III
Efficiency and Market Equilibrium with Scarce Renewable Resources

Chapter 6
Renewable Resources and Intergenerational Efficiency

6.1 Introduction

The attentive reader has probably been somewhat surprised that in a book about resource economics so many chapters have been devoted to the analysis of economic growth with abundant natural resources. However, we will see in the present chapter that economic growth under conditions of utilization of scarce natural resources can in fact be regarded as merely a more complex application of intertemporal allocation and growth theory than that found in standard allocation and growth theory. To avoid premature complication, we first focus in the following chapters on renewable natural resources and leave the problem of economic growth with non-renewable resources in a general equilibrium context to Chap. 8.

There are several reasons why the introduction of renewable natural resources makes growth theory more complex. First, since renewable natural resources survive one model period (in reality 25-30 years), man-made capital is no longer the sole capital good. Farmer and Wendner (2003) show that the intertemporal equilibrium dynamics of OLG models with heterogeneous capital differ substantially from the dynamics of OLG models with homogenous capital. Second, the harvest from the renewable resource represents a further input in the production of produced commodities. Three-factor intertemporal general equilibrium models can thus have different structural characteristics than two-factor models. Third, in contrast to the Diamond model without natural resources, the durability of the renewable resource also means that a further dimension is added to equilibrium dynamics. The analysis of the dynamic stability of higher order-difference equations thus becomes rather more complex.

Keeping all the above in mind, we focus now on the main problem of the economics of renewable resources.[1] The question here is how can an intergenerationally efficient harvest policy be characterized in a growing economy, and which FOCs are typical for efficient harvest policy and efficient asset allocation over time? Clearly,

[1] For a comprehensive survey on the topics addressed in renewable resource economics, see Brown (2000), Conrad and Clark (1987), and Plourde (1970).

K. Farmer and B. Bednar-Friedl, *Intertemporal Resource Economics: An Introduction to the Overlapping Generations Approach*, DOI 10.1007/978-3-642-13229-2_6,

the stock of renewable resources can increase if the harvest rate is less than the natural growth rate. However, the latter is not limitless since it depends on the carrying capacity of nature. Within the carrying capacity of nature renewable resources (referred to formally as 'sustained' resource use) offer the opportunity of infinitely lasting sustainable resource use and thus allow for satisfaction of human needs over and above those which could be met by exhaustible resources.

Renewable resources, however, can still to be exhausted. Resource exhaustion occurs if the harvest rate permanently exceeds the natural growth rate and/or if the ecological habitat of a species is disturbed or even destroyed by human intervention, for example as a result of excessive economic activity.

This brings us to the main topic of this chapter, namely the complex interactions between nature and the economy. The growth of nature provides natural inputs for the production process which in turn allows for accumulation of man-made capital. To be able to investigate the nature-economy interactions analytically we have to specify the growth function of nature. In order to confine ourselves to an acceptable level of mathematics for the present text, we introduce two simplifying assumptions. First, we do not consider ecological interdependence between several species and focus only on a single species taken to be representative of the natural system as a whole.[2] Second, economic-ecological interdependence is modeled in a very rudimentary fashion. Despite these simplifications, the mathematical model still remains rather demanding.

In the following section, the main analytical tool in the economics of renewable resources, the regeneration function, is introduced. Then, the cost function for resource harvesting is specified. In Sect. 6.4 the notion of short-run intergenerationally efficient resource harvest is introduced. The FOCs for short-run intergenerational efficiency are in Sect. 6.5. The final section contains some remarks regarding long-run intergenerational efficiency.

6.2 The Regeneration Function

Use of the so-called 'regeneration function' largely distinguishes the economics of renewable from that of exhaustible resources (Clark, 1990). Renewable resources such as fish stocks, forests, agricultural land, and some environmental media (air, water systems etc.) can all be analyzed by the use of such a function.

Definition 6.1. The regeneration function provides the absolute increase of biomass during a period or at a point in time t measured by number or weight of resource units.

Obviously, the regeneration function itself can vary from simple to complex forms depending on how species growth is measured, e.g. by year of birth, sex or indi-

[2] For multi-species model formulations, see Conrad (1999, chap. 13) for a three level food chain specification, or Clark (1990) for the well-known Lotka-Volterra models of predator-prey dynamics and the dynamics of competition between different species, known as inter-specific competition.

vidual weight. For our general equilibrium analysis, however, we resort to a simple yet realistic form of regeneration function. This type of regeneration corresponds to the sigmoid growth function verified for density dependent species populations (see, e.g., Begon et al, 1996). The basic idea behind this form is that the absolute increase in biomass depends on the magnitude of the existing stock of the resource. For given environmental conditions, i.e. availability of space, light, water, nutrients etc., the increase in the resource rises with the magnitude of the stock: The more brood animals that exist, the more offspring they produce. Nutrient and space limitations are not yet binding. However, with rising resource stock these limitations become binding. As a consequence of this competition between individuals of a species, resource growth becomes smaller with rising stock, and eventually it turns negative. The environmental conditions do not allow for further increases in the resource stock.

These considerations are formalized by the following *logistic* function which we present here in both a continuous-time and a discrete-time version. This function was first used by the biologist Schaefer (1954) for pacific tuna fisheries (Brown, 2000). In the partial equilibrium literature and also in ILA general equilibrium models the continuous-time version (6.1a) normally dominates. However, within our OLG models we need the discrete-time version (6.1b):

$$\frac{\mathrm{d}R(t)}{\mathrm{d}t} = \pi R(t) - \Omega R(t)^2, \; \pi \equiv \Pi - 1 > 0, \; \Omega > 0, \tag{6.1a}$$

$$R_{t+1} \equiv \mathrm{g}(R_t) = \Pi R_t - \Omega R_t^2, \tag{6.1b}$$

where π denotes the growth enhancing and Ω the growth retarding factor due to competition among individuals within a species, e.g. with respect to food, known as intra-specific competition. The difference between Π and π is that the former refers to the growth factor while the latter is the growth rate. As a consequence, there exists a minimum population which coincides with the origin and a maximum viable population, which is called 'carrying capacity' in population ecology models (Begon et al, 1996).

Note that the discrete-time version of the regeneration function follows from the continuous-time version by acknowledging that $\mathrm{d}R(t) \approx R_{t+1} - R_t$ and $\mathrm{d}t \approx 1$.[3] Figure 6.1 presents a graphical illustration of the logistic regeneration function in discrete time.

At point $\underline{R}$ in Figure 6.1 the resource stock is zero.[4] Nonetheless, the reader can see that the slope of the regeneration function which is equal to π is larger than zero by definition. This assumption appears plausible, e.g. for some tree species where

[3] One crucial difference exits however between the continuous-time and discrete time version: In the case of discrete dynamics, the potential for cyclical and chaotic behavior has to be considered, as shown by May (1974). For a discussion, see Clark (1990, 202) or Conrad (1999, 64).

[4] In more complex growth specifications, $\underline{R}$ need not coincide with the origin but can be associated with a positive level of the resource stock. This implies that at a very small but positive level of the resource stock, resource growth becomes negative e.g. as a lack of suitable mates. This phenomenon is called 'critical depensation' (see, e.g. Hanley et al, 2007, 268).

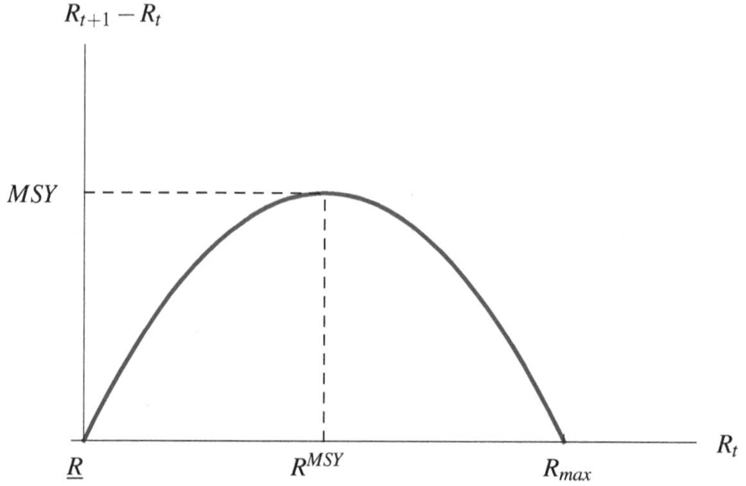

Fig. 6.1 Logistic regeneration function

roots provide new growth for trees, or for fish species where eggs laid in the water may generate new offspring.

From point $\underline{R}$ up to point R^{MSY} along the regeneration function the absolute increase of the resource stock between time t and $t+1$, namely $R_{t+1} - R_t$, increases with rising stock R_t. Environmental constraints are not yet binding and more brood animals generate more offspring.

At point R^{MSY}, where MSY stands for maximum sustainable yield (see Section 6.2.2), the increase in the resource stock over time is maximized.

Along the curve between point R^{MSY} and point R_{max} the resource increase over time is larger than zero but decreasing with rising resource stock, since environmental constraints such as available space and nutrients become binding. The slope of the regeneration function is thus negative.

At point R_{max} the resource stock itself is maximized. The increase in the resource stock is nil. The resource stock has reached the limit of ecological carrying capacity. This brings us to the notion of the natural equilibrium.

6.2.1 The Natural Equilibrium

Definition 6.2. The resource stock R_{max} at which the increase in the resource stock is nil is called the natural equilibrium.

Or more formally for the continuous-time and the discrete-time growth function, respectively:

$$\frac{dR(t)}{dt} = 0, \tag{6.2a}$$

$$R_{t+1} = R_t. \tag{6.2b}$$

On the basis of the logistic regeneration function, it is very easy to see that the natural equilibrium resource stock is as follows:

$$R_{max} = \frac{\pi}{\Omega}. \tag{6.3}$$

It is also easy to see why the term natural equilibrium is warranted. Suppose that the initial resource R_0 differs from R_{max}. Will the natural dynamics depicted by the discrete-time version of the logistic regeneration function automatically bring about a move from R_0 towards R_{max}? The answer is 'yes', since the derivative of the resource stock in period $t+1$ with respect to the resource stock in t, dR_{t+1}/dR_t, is equal to $\Pi - 2\Omega R_t$. Evaluated at the natural equilibrium R_{max}, the same derivative is equal to $\Pi - 2\Omega R_t(\pi/\Omega) = 2 - \Pi < 1$, while at $R = 0$ it is equal to $\Pi > 1$. Thus, the resource dynamics is unstable in the neighborhood of $R = 0$ while the natural equilibrium stock R_{max} represents the single stable fixed point of the logistic resource stock dynamics. In other words: Without human intervention the resource stock will automatically tend towards its natural equilibrium value, i.e. the carrying capacity of the resource stock.

6.2.2 The Sustainable Yield

While without human intervention the (logistic) resource dynamics converges towards the natural equilibrium, humans cannot survive without harvesting X_t of the natural resource R_t. Such harvesting can be formally represented in a net regeneration function as follows, again in the continuous-time and in the discrete-time version, respectively:

$$\frac{dR(t)}{dt} = \pi R(t) - \Omega R(t)^2 - X(t), \tag{6.4a}$$

$$R_{t+1} = \Pi R_t - \Omega R_t^2 - X_t \equiv g(R_t) - X_t. \tag{6.4b}$$

With the aid of the net regeneration function we are able to introduce the notion of 'sustainable yield' or 'sustainable harvest volume'.

Definition 6.3. The 'sustainable yield' ensures that in spite of human intervention (harvesting) the resource stock does not change over time. The harvest volume equals natural regeneration.

Or more formally:

$$X = \pi R - \Omega R^2. \tag{6.5}$$

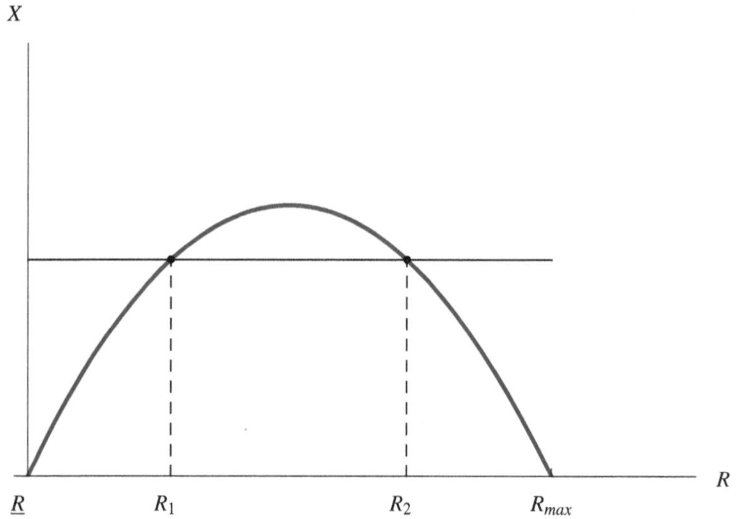

Fig. 6.2 Emergence of two bio-economic equilibria R_1 and R_2

It is easy to see that for any given X (6.5) represents a quadratic equation in R which allows for exactly two solutions. All points on the net regeneration function, i.e. combinations of harvest X and resource stock R that fulfill

$$\{(X,R)\,|X = \pi R - \Omega R^2\}$$

are called bio-economic equilibria of logistic natural regeneration. In general, two solutions emerge as depicted by R_1 and R_2 in Figure 6.2. However, a unique solution results as *maximum sustainable yield* (MSY) level:

$$\max X \Leftrightarrow \frac{\mathrm{d}X}{\mathrm{d}R} = 0 \Rightarrow R = \frac{\pi}{2\Omega}. \tag{6.6}$$

Hence, the resource stock at which maximum sustainable yield occurs equals exactly half of the natural equilibrium stock, i.e. $R^{MSY} = R_{max}/2$. This is typical for logistic regeneration in which the regeneration function is symmetric about the MSY resource stock.

6.2.3 The Own Rate of Return

For all resource stocks smaller than the resource stock associated with maximum sustainable yield (R^{MSY}) the slope of the regeneration function is larger than zero, while for larger resource stocks up to the natural equilibrium stock R_{max} the slope is negative. Thus, the question arises as to what economic interpretation can be attributed to the slope of the regeneration function. The answer is provided by the notion of the *own rate of return* on the renewable resource, now defined below.

Definition 6.4. The own rate of return on the renewable resource gives the resource increase or the change of the sustainable yield (harvest) for a marginal change in the resource stock. It equals the slope of the net regeneration function.

Looking at the logistic regeneration function we immediately see that the own rate of return on the renewable resource is larger than zero (positive) only within the range of resource stocks which are smaller than the *MSY* resource stock. For a larger resource stock the own rate of return on the renewable resource is negative. At the *MSY* resource stock the own rate of return is nil. It is important to remember these properties of the own rate of return for a logistically regenerating natural resource because the natural resource, as an asset, has to compete with man-made capital by providing an own rate of return as least as large as the real interest rate (= real rate of return on man-made capital). Thus, we will find at the end of this chapter that for any given positive real interest rate the resource stock needs to be harvested in such a way that its own rate of return is positive and equal to the real interest rate.

6.3 The Harvest Cost Function

For the sake of simplicity, in the resource economics literature, namely general equilibrium models, zero or constant average harvest costs are often assumed (Berck, 1981). This is clearly not a realistic assumption and this is why resource harvest costs are commonly found in sectoral models (for an overview, see Clark, 1990; Neher, 1990; Brown, 2000). It is a simple fact that renewable resources cannot be harvested without the use of scarce factors (labor, man-made capital), and we thus find, as with regeneration functions, simple or complex forms of harvest cost functions. A functional specification popular in fishery models is again the Schaefer (1954) version where a fixed amount of effort per resource stock unit, known as catchability constant, is needed to harvest a unit of the stock. A problem with this specification is that the resulting harvest production function exhibits increasing returns to scale (Conrad, 1999), and this is why we use an alternative specification in the remainder of this chapter.

The simplest specification of a harvest cost function assumes that (i) only human labor is necessary to harvest the resource and (ii) that labor, measured in hours, increases progressively with rising harvest volume. The following quadratic functional specification is in line with these properties.

$$N_t^R = N^R(X_t) = \left(\frac{1}{2}\right) \mathrm{A}(X_t)^2, \tag{6.7}$$

where A is a positive constant. However, the hours the agents work in resource harvesting cannot simultaneously be spent in production. Resource harvesting thus competes with commodity production and this is formally expressed by the following full employment condition:

$$N_t^Y + N_t^R = L_t, \tag{6.8}$$

where N_t^Y and N_t^R stand for labor demanded for commodity production and for resource harvesting, respectively. Now that we have described the main new complexities concerning the problem of intergenerational efficiency in economic growth with renewable resources, we are ready to state the problem itself in more exact terms. This is dealt with in the next section.

6.4 Short-Run Intergenerationally Efficient Resource Harvesting

We know from Chap. 3 that an intertemporal market equilibrium with abundant natural resources is short-run intergenerationally efficient. Thus, it is natural to suggest that this is also true if a renewable resource is used both as an production input and as an alternative asset. In order to verify whether this is true or not we first need to define the nature of a short-run intergenerationally efficient consumption allocation. To that end, we set up below the multi-period optimization problem of the initial older generation. To simplify the analysis, we assume that $G^{\tau} = 1$, i.e. the rate of technological progress assumed is zero, and the growth factor of population growth $G^L = 1$, i.e. the population growth rate is zero. Thus, the production function with resource harvest X_t, labor N_t^Y and man-made capital K_t reads as follows:

$$Y_t \equiv \mathrm{F}(X_t, N_t^Y, K_t) = (X_t)^{\alpha_1} \left(N_t^Y\right)^{\alpha_2} (K_t)^{\alpha_3}.$$

We know from Chap. 3 that short-run intergenerational efficiency is equivalent to the solution of the following multi-period optimization problem for the initial older generation:

$$\max \rightarrow \ln c_{-1}^1 + \beta \ln c_0^2$$

subject to the constraints:

$$\ln c_t^1 + \beta \ln c_{t+1}^2 \geq \left(U_t^1\right)^{\circ}, \; t = 0, 1, ..., T-1, \tag{6.9a}$$

$$L_t c_t^1 + L_{t-1} c_t^2 + K_{t+1} - K_t \leq (X_t)^{\alpha_1} \left(N_t^Y\right)^{\alpha_2} (K_t)^{\alpha_3} - \delta K_t, \qquad \alpha_1 + \alpha_2 + \alpha_3 = 1, \; \forall t, \tag{6.9b}$$

$$N_t^Y + N_t^R \leq L_t = L_0, \tag{6.9c}$$

$$N_t^R = N^R(X_t) = \left(\frac{1}{2}\right) \mathrm{A} (X_t)^2, \tag{6.9d}$$

$$R_t = \Pi R_{t-1} - \Omega R_{t-1}^2 - X_{t-1}, \tag{6.9e}$$

$$c_T^1 \geq \underline{c}^1, \; K_{T+1} \geq \underline{K}, \; R_T \geq \underline{R}. \tag{6.9f}$$

The objective function and the first set of constraints (6.9a) are the same as in Chap. 3. The next constraint, (6.9b), consists of the aggregate consumption and investment constraint, into which the aggregate production function has already been inserted. The main difference in comparison to chapter three is the inclusion of the resource harvest as an input into commodity production. Constraint (6.9c) ensures that employment in resource harvesting and in commodity production is not larger than the labor force. Then, the harvest cost function and the net regeneration function have to be respected as further constraints, see (6.9d)–(6.9e). Finally, the consumption of the younger generation as well as the stocks of man-made capital and of the renewable resource at the end of the optimization horizon must be as large as some pre-specified minimal levels which ensure that the economy is able to move on after period T. Last but not least it is assumed that c^1_{-1},K_0, R_{-1}, and L_{-1} are exogenously fixed.

As also known from Chap. 3, the first step in solving this optimization problem is to set up the corresponding Lagrangian. Recall that the Lagrangian multipliers, with the exception of the constraints for younger households' utility functions, are specified in non-discounted form. Hence, for example ϕ^L_t denotes the non-discounted multiplier of the employment constraint while $\phi^L_t \beta^t$ is the corresponding discounted multiplier.

The Lagrangian associated with the above optimization problem reads as follows:

$$\begin{aligned}
\mathscr{L} &= \ln c^1_{-1} + \beta \ln c^2_0 + \\
&+ \sum_{t=0}^{T-1} \mu^c_t \left[\ln c^1_t + \beta \ln c^2_{t+1} - \left(U^1_t\right)^{\circ} \right] + \sum_{t=0}^{T} \phi^L_t \beta^t \left[L_t - N^Y_t - (1/2)\mathrm{A}\left(X_t\right)^2 \right] + \\
&+ \sum_{t=0}^{T} \phi^Y_t \beta^t \left[(X_t)^{\alpha_1} \left(N^Y_t\right)^{\alpha_2} (K_t)^{\alpha_3} + (1-\delta) K_t - L_t c^1_t - L_{t-1} c^2_t - K_{t+1} \right] + \\
&+ \sum_{t=0}^{T} \phi^R_t \beta^t \left[\Pi R_{t-1} - \Omega R^2_{t-1} - X_{t-1} - R_t \right] + (\mu^c_T) \beta^T \left[\underline{c}^1 - c^1_T \right] + \\
&+ \phi^K_{T+1} \beta^{T+1} \left[\underline{K} - K_{T+1} \right] + \phi^R_T \beta^T \left[\underline{R} - R_T \right].
\end{aligned} \tag{6.10}$$

6.5 FOCs for Short-Run Intergenerational Efficiency

As usual, the FOCs for short-run intergenerational efficiency are obtained by differentiating the Lagrangian above (6.10) with respect to all consumption quantities, production inputs, stock variables and all Lagrangian multipliers, and setting the derivatives equal to zero. We thus obtain the following FOCs which can then be interpreted economically. Here we find that the non-discounted Lagrangian multipliers play the role of so-called 'shadow prices'. We start with the FOC for efficient allocation of retirement consumption of old households alive in period t:

$$MRS \equiv \beta/(c_0{}^2)^\circ = (\phi_0{}^Y)^\circ L_{-1}. \tag{6.11}$$

The interpretation of this FOC is the same as in Chap. 3.

For the households from period t onwards, consumption has to be allocated between working and retirement period according to:

$$-\frac{\mathrm{d}c_{t+1}^2}{\mathrm{d}c_t^1} \equiv \underbrace{\frac{\left(c_{t+1}^2\right)^\circ}{\beta\left(c_t^1\right)^\circ}}_{\frac{\partial U_t^1/\partial c_t^1}{\partial U_t^1/\partial c_{t+1}^2}} = \frac{\left(\phi_t^Y\right)^\circ(1+\theta)}{\left(\phi_{t+1}^Y\right)^\circ}, \tag{6.12}$$

where $\beta \equiv (1/(1+\theta))$. This FOC concerns the intertemporal optimal consumption bundle for the working and retirement period of period-t younger household. To interpret the FOC economically, note that ϕ_t^Y indicates the utility value of one marginally larger unit of production in period t. Acknowledging this interpretation of ϕ_t^Y, the above FOC says that the intertemporal marginal rate of substitution between present and future consumption of period-t younger generation which is identical to the ratio of marginal utilities of present and future consumption has to be equal to the ratio of the present to the discounted future shadow price of production. As already known, the intertemporal marginal rate of substitution between present and future consumption tells us how many consumption units in the retirement period the younger household is ready to forego in order to obtain one additional present consumption unit. The ratio of the present to the discounted future shadow price of production equals the number of future production units the economy has to sacrifice if one additional production unit is used for present consumption. Or in other words: Intertemporal consumption efficiency demands that the younger household is ready to sacrifice exactly the same number of consumption units in the retirement period for one additional consumption unit in the working period as the number of future production (and consumption) units the economy has to sacrifice for one additional unit in the current period.

The first order condition for efficient allocation of labor requires:

$$\underbrace{\left(\phi_t^Y\right)^\circ(1-\alpha_1-\alpha_3)\left(\frac{(X_t)^\circ}{\left(N_t^Y\right)^\circ}\right)^{\alpha_1}\left(\frac{(K_t)^\circ}{\left(N_t^Y\right)^\circ}\right)^{\alpha_3}}_{\partial \mathrm{F}/\partial N_t^Y} = \left(\phi_t^L\right)^\circ. \tag{6.13}$$

The left hand side of (6.13) gives the marginal value product of labor in commodity production (in utility units), and the right hand side gives the shadow price of the labor force, i.e the utility of a marginally higher unit of labor force. The latter equals the (marginal) opportunity costs (in utility units) of one additional employee in commodity production. Acknowledging these interpretations, short-run intergenerational efficiency demands that an additional employee in commodity production generates exactly the same additional utility as he or she costs, i.e. the additional utility which an employee would generate in resource harvesting.

The first order condition for efficient resource harvesting requires that the marginal value product of one additionally harvested resource unit equals the marginal harvest costs plus the marginal opportunity costs of an additionally harvested resource unit, these latter being equal to the utility of a resource unit not harvested:

$$\underbrace{\left(\phi_t^Y\right)^\circ \alpha_1 \left(\frac{(X_t)^\circ}{\left(N_t^Y\right)^\circ}\right)^{\alpha_1 - 1} \left(\frac{(K_t)^\circ}{\left(N_t^Y\right)^\circ}\right)^{\alpha_3}}_{\partial \mathrm{F}/\partial X_t} = \left(\phi_t^L\right)^\circ \underbrace{\mathrm{A}X_t}_{\mathrm{N}^{R'}(X_t)} + \frac{\left(\phi_{t+1}^R\right)^\circ}{(1+\theta)}. \tag{6.14}$$

Thus, on the left hand side of (6.14) stands the marginal value product of the resource harvest (in utility units), while the right hand side is composed of the marginal harvest costs and the marginal user costs of an additional resource unit harvested and not conserved. $\left(\phi_{t+1}^R\right)^\circ/(1+\theta)$ is the discounted marginal value (in utility units) of one conserved (unharvested) resource unit.

The short-run intergenerationally efficient accumulation of the renewable resource stock is governed by the following first order condition:

$$\left(\phi_{t+1}^R\right)^\circ \underbrace{\left[\Pi - 2\Omega\,(R_t)^\circ\right]}_{g'((R_t)^\circ)} = (1+\theta)\left(\phi_t^R\right)^\circ. \tag{6.15}$$

The left hand side of (6.15) represents the shadow price of an additional resource unit at the beginning of period $t+1$ times the own factor of return on the renewable resource at the beginning of period $t+1$. The right hand side shows the marginal value of postponing resource harvest by one period. The latter is equal to the marginal opportunity of a resource unit not harvested but left to augment the resource stock. Thus, intergenerational efficiency demands that the utility increase from an additional resource unit not harvested but left to increase the resource, equals the marginal opportunity costs of the resource unit not harvested, i.e. the utility value of an in period t additionally harvested resource times the subjective time preference factor.

$$\left(\phi_{t+1}^Y\right)^\circ \left[1 - \delta + \underbrace{\alpha_3 \left(\frac{(X_{t+1})^\circ}{\left(N_{t+1}^Y\right)^\circ}\right)^{\alpha_1} \left(\frac{(K_{t+1})^\circ}{\left(N_{t+1}^Y\right)^\circ}\right)^{\alpha_3 - 1}}_{\partial F/\partial K_{t+1}}\right] = (1+\theta)\left(\phi_t^Y\right)^\circ \tag{6.16}$$

On the left hand side of this equality stands the utility value of an additional unit of production times the own return factor on man-made capital between t and $t+1$, and on the right hand side we find the utility value of an additional unit of production in period t times the subjective time preference factor. Intergenerational efficiency demands that the utility increase of an additional unit of man-made capital at the beginning of $t+1$ exactly equals the marginal opportunity costs of a commodity unit invested in man-made capital, which are equal to the utility of an additional unit of production not invested in t, times the time preference factor.

Clearly, besides the FOCs interpreted so far, there are also additional Lagrangian constraints. When the corresponding Lagrangian multipliers are nonzero, these additional constraints hold as equalities and an interior solution exists.

6.6 Long-Run Intergenerational Efficiency

Having thus characterized short-run intergenerational efficiency in terms of the utility maximization of the initial older generation subject to a finite series of constraints, we now focus in this section on long-run intergenerational efficiency. Notice that in the following we do not seek to determine whether a steady state solution exists, nor whether it is dynamically stable. In this section we merely assume that this is the case for the OLG model with renewable resources introduced above. We leave rigorous analysis of these problems to Chaps. 9 and 10.

As in Chap. 5, we permutate the role of the older and younger household within the long-run efficiency calculus: the life-cycle utility of the younger household is maximized subject to the constraints that the initial older generation obtains a certain utility level and the constraints (6.9b)–(6.9f) in steady state hold:

$$
\begin{aligned}
&\max \quad \ln c^1 + \beta \ln c^2 \\
&\text{subject to:} \\
&\ln c_0^2 \geq \ln(c_0^2)^{\bullet}, \\
&L_0 c^1 + L_0 c_0^2 \leq (X)^{\alpha_1} \left(N^Y\right)^{\alpha_2} (K_0)^{\alpha_3} - K, \\
&L_0 c^1 + L_0 c^2 \leq (X)^{\alpha_1} \left(N^Y\right)^{\alpha_2} (K)^{\alpha_3} - K, \\
&N^Y + N^R \leq L = L_0, \\
&N^R = (1/2)\mathrm{A}(X)^2, \\
&\mathrm{g}(R) = X, \\
&R_0 + \mathrm{g}(R_0) = R + X.
\end{aligned}
$$

Setting up the Lagrangian

$$
\begin{aligned}
\mathscr{L} = {} & \ln c^1 + \beta \ln c^2 + \mu_{-1}^c [\ln c_0^2 - \ln(c_0^2)^{\bullet}] + \phi_0^y \left[(X)^{\alpha_1} \left(N^Y\right)^{\alpha_2} (K_0)^{\alpha_3} - \right. \\
& \left. - K - L_0 c^1 - L_0 c_0^2\right] + \phi^y \left[(X)^{\alpha_1} \left(N^Y\right)^{\alpha_2} (K)^{\alpha_3} - K - L_0 c^1 - L_0 c^2\right] + \\
& + \phi^L \left[L_0 - N^Y - (1/2)\mathrm{A}(X)^2\right] + \phi^R [\mathrm{g}(R) - X] + \phi_0^R [R_0 + \mathrm{g}(R_0) - X - R],
\end{aligned}
$$

yields the following first order conditions:

$$\frac{(c^2)^\bullet}{\beta(c^1)^\bullet} = 1 + \frac{(\phi_0^y)^\bullet}{(\phi^y)^\bullet}, \tag{6.17}$$

$$\frac{(\mu_{-1}^c)^\bullet}{(c_0^2)^\bullet} = (\phi_0^y)^\bullet L_0, \tag{6.18}$$

$$\alpha_3 \frac{Y^\bullet}{K^\bullet} = 1 + \frac{(\phi_0^y)^\bullet}{(\phi^y)^\bullet}, \tag{6.19}$$

$$(\phi_0^y)^\bullet \alpha_1 \frac{(Y_0)^\bullet}{X^\bullet} + (\phi^y)^\bullet \alpha_1 \frac{Y^\bullet}{X^\bullet} = (\phi^L)^\bullet \mathrm{A} X^\bullet + (\phi^R)^\bullet + (\phi_0^R)^\bullet, \tag{6.20}$$

$$(\phi_0^y)^\bullet \alpha_2 \frac{(Y_0)^\bullet}{(N^y)^\bullet} + (\phi^y)^\bullet \alpha_2 \frac{Y^\bullet}{(N^y)^\bullet} = (\phi^L)^\bullet, \tag{6.21}$$

$$(\phi^R)^\bullet g'(R^\bullet) = (\phi_0^R)^\bullet. \tag{6.22}$$

Proposition 6.1 characterizes the properties of long-run intergenerationally efficient and inefficient steady states.

Proposition 6.1 (Long-run intergenerational efficiency). *A long-run intergenerationally efficient consumption allocation* $\{(c^1)^\bullet, (c^2)^\bullet, c_0^2)^\bullet\}$ *associated with steady state man-made capital* $K^\bullet$*, natural resource stock* $R^\bullet$*, resource harvest* $X^\bullet$*, and employment in production* $(N^y)^\bullet$*, can be characterized as follows:* $(c^2)^\bullet/(\beta(c^1)^\bullet) = 1 + (\phi_0^y)^\bullet/(\phi^y)^\bullet \geq 1$, $(\mu_{-1}^c)^\bullet/(c_0^2)^\bullet = (\phi_0^y)^\bullet L_0$, $\alpha_3 Y^\bullet/K^\bullet = 1 + (\phi_0^y)^\bullet/(\phi^y)^\bullet \geq 1$, $(\phi^R)^\bullet g'(R^\bullet) > 0$, $X^\bullet = g(R^\bullet)$, $(\phi_0^y)^\bullet \alpha_1 (Y_0)^\bullet/$ $X^\bullet + (\phi^y)^\bullet \alpha_1 Y^\bullet/X^\bullet = (\phi^L)^\bullet \mathrm{A} X^\bullet + (\phi^R)^\bullet$, $(\phi^y)^\bullet \alpha_2 Y^\bullet/(N^y)^\bullet = (\phi^L)^\bullet$.

The golden age allocation with Golden rule K^{GR} *and* R^{MSY} *is characterized as follows:* $(c^2)^\bullet/(\beta(c^1)^\bullet) = 1 = \alpha_3 Y^\bullet/K^{GR}$, $g'(R^{MSY}) = 0$, $X^{MSY} = g(R^{MSY})$, $(\phi^y)^\bullet \alpha_1 Y^\bullet/X^{MSY} = (\phi^L)^\bullet \mathrm{A} X^{MSY} + (\phi^R)^\bullet$, $(\phi^y)^\bullet \alpha_2 Y^\bullet/(N^y)^\bullet = (\phi^L)^\bullet$.

Several comments regarding Proposition 6.1 are in order. First, long-run intergenerational efficiency in general demands that the resource harvest $X^\bullet$ and the associated resource stock are lower than maximum sustainable yield (and the associated resource stock). Moreover, the man-made capital stock is lower than the Golden rule capital stock.

Second, maximum sustainable yield emerges only in a golden age path which presupposes that $(\mu_{-1}^c)^\bullet = 0$ and $(\phi_0^y)^\bullet = (\phi_0^R)^\bullet = 0$. Moreover, in golden age the intertemporal marginal rate of substitution between young and old consumption equals one which is also true for the marginal productivity of man-made capital.

Figure 6.3 illustrates the main long-run efficiency conditions in Prop. 6.1. The condition regarding the efficient resource harvest volumes is illustrated by the upper part of the diagram in which resource harvest volumes are depicted on the abscissa, while on the vertical axis the marginal value product and the marginal harvest costs (in utility units) of alternative harvest quantities are depicted. The diagram also shows that the user costs of resource harvest, $(\phi^R)^\bullet(1 + g'(R^\bullet)$ drive a wedge

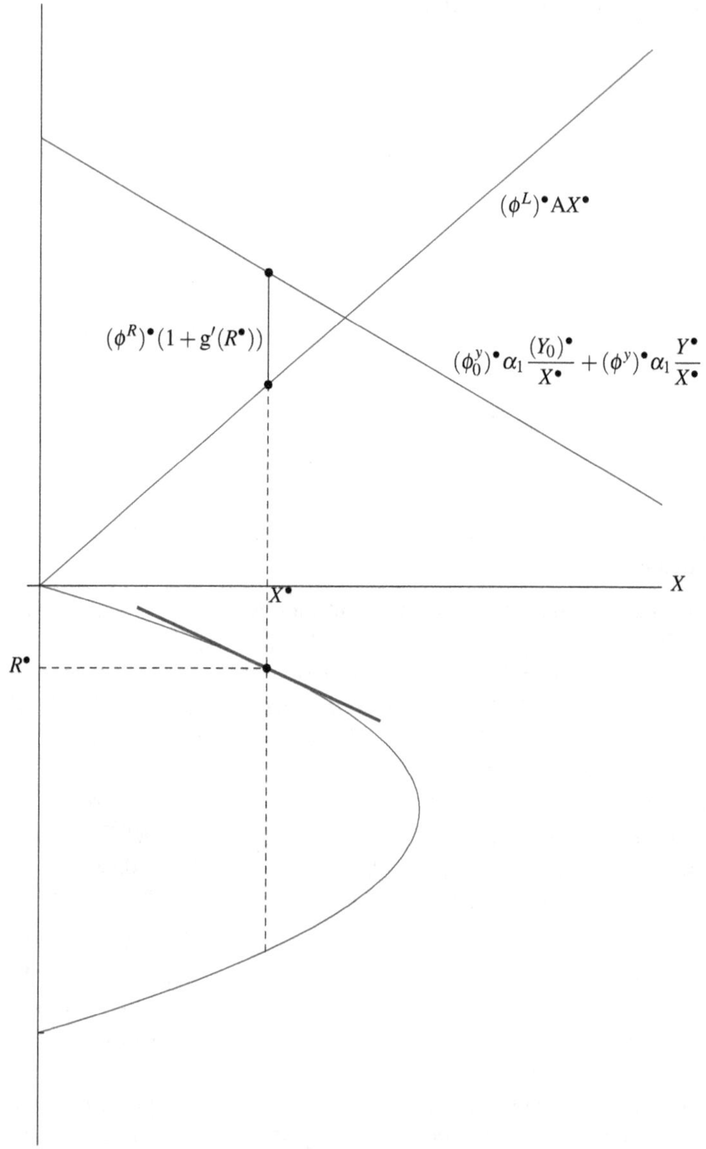

Fig. 6.3 Long-run intergenerational efficient resource harvest and resource stock for $(\mu_{-1}^c)^\bullet > 0$.

between the marginal value product and the marginal harvest cost such that the long-run efficient harvest volume is less than that which follows from the intersection of marginal value product and marginal harvest cost curve. The lower part of the diagram depicts the net regeneration function with resource stocks on the downward pointing ordinate, and again harvest volumes on the abscissa. At the long-run intergenerationally efficient harvest volume $X^{\bullet}$ and the corresponding resource stock $R^{\bullet}$, the own rate of return on the renewable resource is shown by the slope of the tangent to the net regeneration function and it is larger than zero for $(\mu_{-1}^{c})^{\bullet} > 0$.

Whether the long-run intergenerationally efficient solution and the bio-economic optimal solution (maximum sustainable yield) are equivalent depends on whether the interests of the initially old generation are taken into account or not. If the interests of the oldest generation matters, it is not long-run intergenerationally efficient to choose *MSY* and the corresponding resource stock. Moreover, with decreasing marginal value product and increasing marginal harvest cost of alternative harvest volumes the long-run intergenerationally efficient harvest volume and the corresponding resource stock are smaller than the bio-economic optimal harvest volume and resource stock. Thus, while the long-run intergenerationally efficient solution can coincide with the bio-economic optimal solution on the golden age path, these will in general not be met and the long-run intergenerationally efficient solution deviates from that which were optimal from an bio-economic perspective.

6.7 Conclusions

In this chapter the major analytical instrument of renewable resource economics, namely the regeneration function, and associated concepts like maximum sustainable yield were introduced into the efficiency calculus of Chap. 3. We derived and interpreted the FOCs for short-run intergenerationally efficient paths for resource harvest, resource stock, man-made capital and the consumption allocation between younger and older households. We also found that long-run intergenerationally efficient harvest volumes and stock levels are smaller than bio-economic optimal levels if the interests of the initially older generation are included. This result leaves open the question whether an intertemporal market equilibrium allocation qualifies for short-run and/or long-run intergenerational efficiency—a question which is dealt with in the following chapter.

References

Begon M, Harper JL, Townsend CR (1996) Ecology. Individuals, Populations and Communities, 3rd edn. Blackwell, Malden, MA

Berck P (1981) Optimal management of renewable resources with growing demand and stock externalities. Journal of Environmental Economics and Management 8(2):105–117

Brown GM (2000) Renewable natural resource management and use without markets. Journal of Economic Literature 38(4):875–914

Clark CW (1990) Mathematical Bioeconomics: The Optimal Management of Renewable Resources, 2nd edn. Wiley, NY

Conrad JM (1999) Resource Economics. Cambridge University Press, Cambridge, New York

Conrad JM, Clark CW (1987) Natural Resource Economics: Notes and Problems. Cambridge University Press, Cambridge, New York

Farmer K, Wendner R (2003) A two–sector overlapping generations model with heterogenous capital. Economic Theory 22:773–792

Hanley N, Shogren JF, White B (2007) Environmental Economics in Theory and Practice. Oxford University Press, Cambridge, Oxford

May R (1974) Biological populations with non-overlapping generations: stable points, stable cycle and chaos. Science 186:645–647

Neher P (1990) Natural Resource Economics: Conservation and exploitation. Cambridge University Press, Cambridge, New York

Plourde CG (1970) A simple model of replenishable natural resource exploitation. American Economic Review 68:518–522

Schaefer MB (1954) Some aspects of the dynamics of populations important to the management of commercial marine fisheries. Bulletin of Inter–American Tropical Tuna Commission 1:25–56

Chapter 7
Intertemporal Market Equilibrium and Intergenerational Efficiency with Renewable Resources

7.1 Introduction

It is widely held that self–interest under the pressure of market competition leads individuals to overexploit and even exhaust renewable resources world-wide. We know however from Chap. 1 that the empirical evidence, at least for non–renewable resources, is much more mixed, i.e. does not generally support this claim. It is the main objective of this chapter to show that theoretical considerations also suggest that we need to exercise considerable caution and judgement when making claims of this sort. Indeed, it would appear that theoretical support is only forthcoming here when no suitable property right regime exists for the renewable resource. Hence, the empirically observed misuse of natural resources is not per se the result of individual self interest and market competition. On the other hand, for many renewable resources it is often hard in practice to define and implement property rights capable of preventing self–interested individuals, acting under conditions of market competition, from overusing renewable resources.

To delineate these general considerations within the theoretical framework we have set up so far, we first introduce a renewable resource into the intertemporal market equilibrium model (from Chap. 4). We consider below two property rights regimes: one where the resource stock is privately owned by the households, and where it is common property and there is open access to resource harvest. Then, we show why under these alternative property rights regimes the market allocation is either short-run intergenerationally efficient or inefficient. Intergenerational efficiency translates also into long-run intergenerational efficiency as will be shown graphically at the end of the chapter.

K. Farmer and B. Bednar-Friedl, *Intertemporal Resource Economics: An Introduction to the Overlapping Generations Approach*, DOI 10.1007/978-3-642-13229-2_7,

7.2 Intertemporal Market Equilibrium with a Privately Owned Resource

Private ownership of the renewable resource presupposes certain economic conditions, i.e. the economic value of harvest plus stock needs to be related to the respective costs of defining, implementing and controlling property rights in the resource. If such costs are lower than economic value, a private property rights regime will emerge. Since the renewable resources under consideration are durable, the private property rights must be such as to extend over the full lifetime of the resource.

The costs of defining and implementing private property rights vary considerably depending on the type of resource under examination. Property rights to immobile renewable resources such as forests are easier to define than property rights to birds or marine fish. The costs for the definition and control of exclusive private property rights to herring ponds or mushroom cultivation are relatively low, and hence these renewable resources are good examples of the kind of resources we have in mind in this chapter.

Exclusive private property rights to the renewable resource are necessary for the emergence of trade. Trade in the resource can mean either that the resource stock itself is the subject of trade, or that there is only trade in resource harvest. Given the demographic structure of the OLG model in Chap. 4, we know that trade in the resource stock typically occurs between the older and the younger generation in any market period, while resource harvest is typically traded between the owner of the resource stock and the producers who need the resource harvest for commodity production.

As in Chap. 4 we assume that the existence of well organized markets for man-made capital, labor and the produced commodity, i.e. that perfectly competitive suppliers and demanders interact in current and in future market periods. In this chapter we extend these assumptions to include markets for the resource stock and the resource harvest. In particular, we assume that at the beginning of each period the resource stock is inelastically supplied to the resource stock market by the older household, while the younger household demands the resource stock at the start of the period in order to be able to harvest the resource during the working period and to sell the harvest yield to the firm on a so-called 'one-period forward' market, where the price of the resource harvested during the period is negotiated at the start of the period (see Foley, 1975; Farmer, 1989, for more details). The 'spot' price of the resource stock at the beginning of period t (for delivery in the same period) is denoted by p_t while the one-period forward price of the resource harvest in period t is denoted by q_t. Notice that both prices are relative prices indicating units of produced commodity per unit of the resource stock and resource harvest respectively.

7.2.1 Individual Optimization Problems

As in Chap. 4, the production sector maximizes one-period real profits defined as the difference between output $Y_t \equiv \mathrm{F}(X_t, N_t^Y, K_t) = (X_t)^{\alpha_1} \left(N_t^Y\right)^{\alpha_2} (K_t)^{\alpha_3}$ and real factor costs which in addition to labor and capital costs now also consist of the costs for the resource harvest. For the sake of simplicity, we set $g^L = 1, \forall t$ (no population growth). We thus have:

$$\max\,(X_t)^{\alpha_1} \left(N_t^Y\right)^{\alpha_2} (K_t)^{\alpha_3} - w_t N_t^Y - r_t K_t - q_t X_t. \tag{7.1}$$

The optimization problem of the initially older household (in $t = 0$) is, as in the basic model without resources, trivial. The representative initially older household receives revenues from sales of the man-made capital stock (in per capita terms). Moreover, since we assume in this section that property rights over the renewable resource are well defined, the household also receives revenues from sale of the resource stock. This gives:

$$\max \beta \ln c_0^2 \tag{7.2}$$

subject to:

$$c_0^2 = r_0 \frac{K_0}{L_{-1}} + p_0 \frac{R_0}{L_{-1}}. \tag{7.3}$$

The younger household has the same objective function as in the basic model (lifetime utility depends on consumption in working and retirement periods) but both the budget constraint for the working and the retirement period differ from the corresponding constraints in the basic model. The budget constraint for the working period includes the spot purchases of the resource stock as expenses and the one–period forward sales of the resource harvest as revenues. When young, the household receives wage income for the share of labor which is devoted to commodity production (see the term in square brackets in (7.5a)), and the revenues from the selling of the resource harvest to the production sector. In the retirement period, the representative household receives revenues from the spot sales of the man-made capital stock and the resource stock. Since the resource is privately owned by the younger household, i.e. the household has exclusive decision rights concerning resource accumulation and harvesting, the regeneration function is included among the constraints of the utility maximization problem. This leaves us with:

$$\max \ln c_t^1 + \beta \ln c_{t+1}^2 \tag{7.4}$$

subject to the constraints:

$$c_t^1 + \frac{K_{t+1}}{L_t} + p_t \frac{R_t}{L_t} = w_t \left[1 - \frac{(1/2)\mathrm{A}(X_t)^2}{L_t}\right] + q_t \frac{X_t}{L_t}, \tag{7.5a}$$

$$c_{t+1}^2 = r_{t+1} \frac{K_{t+1}}{L_t} + p_{t+1} \frac{R_{t+1}}{L_t}, \tag{7.5b}$$

$$R_{t+1} = \Pi R_t - \Omega \left(R_t\right)^2 - X_t. \tag{7.5c}$$

7.2.2 First Order Conditions for Individual Maxima

The FOC which determines profit maximizing labor input N_t^Y in commodity production reads as follows:

$$(1-\alpha_1-\alpha_3)\left(\frac{X_t^d}{N_t^Y}\right)^{\alpha_1}\left(\frac{K_t^d}{N_t^Y}\right)^{\alpha_3}=w_t. \tag{7.6}$$

Profit maximizing harvest input, denoted by X_t^d, is determined as follows:

$$\alpha_1\left(\frac{X_t^d}{N_t^Y}\right)^{\alpha_1-1}\left(\frac{K_t^d}{N_t^Y}\right)^{\alpha_3}=q_t. \tag{7.7}$$

Finally, profit maximizing capital input, denoted by K_t^d, is determined as follows:

$$\alpha_3\left(\frac{X_t^d}{N_t^Y}\right)^{\alpha_1}\left(\frac{K_t^d}{N_t^Y}\right)^{\alpha_3-1}=r_t. \tag{7.8}$$

The optimal consumption of the older household in the initial period comes from the budget constraint:

$$L_{-1}c_0^2=r_0K_0+p_0R_0. \tag{7.9}$$

As in the basic model with resource abundance, the younger household attempts to maximize life–cycle utility and thus equates the intertemporal marginal rate of substitution to the factor of return on man-made capital in period $t+1$:

$$\frac{c_{t+1}^2}{\beta c_t^1}=(1+\theta)\frac{(c_{t+1}^2)}{(c_t^1)}=r_{t+1}. \tag{7.10}$$

Moreover, the Hotelling rule modified for renewable resources requires that the factor of return on the resource $p_{t+1}/p_t(\Pi-2\Omega R_t^d)$ and on man-made capital r_{t+1} balance:

$$p_{t+1}=\frac{r_{t+1}}{\Pi-2\Omega\left(R_t^d\right)}p_t\equiv\frac{r_{t+1}}{1+g'\left(R_t^d\right)}p_t. \tag{7.11}$$

Clearly, in maximizing life-cycle utility the younger household also acknowledges the net regeneration function. Thus:

$$R_{t+1}=\Pi R_t^d-\Omega\left(R_t^d\right)^2-X_t\equiv g\left(R_t^d\right)-X_t. \tag{7.12}$$

Between the one–period forward price of resource harvest and the next period spot price of the resource stock augmented by marginal harvest cost there is another no-arbitrage condition which is presented in the following equality:

$$q_t=\frac{p_{t+1}}{r_{t+1}}+w_t\underbrace{AX_t}_{N^{R'}(X_t)}. \tag{7.13}$$

Inserting the net regeneration function (7.12) into the budget constraint for the retirement period (7.5b), we get:

$$c_{t+1}^2 = r_{t+1}\left(\frac{K_{t+1}}{L_t}\right) + p_{t+1}\left(\frac{\Pi R_t - \Omega\left(R_t^d\right)^2 - X_t}{L_t}\right).$$

Solving this equation for K_{t+1}/L_t and inserting this expression for K_{t+1}/L_t into (7.5a) gives:

$$c_t^1 + \frac{c_{t+1}^2}{r_{t+1}} - \frac{p_{t+1}}{r_{t+1}}\frac{\left[\Pi R_t^d - \Omega\left(R_t^d\right)^2\right]}{L_t} + p_t\frac{R_t^d}{L_t} = w_t\left[1 - \left(\frac{(1/2)\mathrm{A}\left(X_t\right)^2}{L_t}\right)\right] + \\ + \left(q_t - \frac{p_{t+1}}{r_{t+1}}\right)\frac{X_t}{L_t}. \quad (7.14)$$

The next step is to take in (7.14) the no-arbitrage conditions (7.11) and (7.13) into account. This yields the intertemporal budget constraint:

$$c_t^1 + \frac{c_{t+1}^2}{r_{t+1}} + \left(\frac{p_{t+1}}{r_{t+1}}\right)\underbrace{\left[\mathrm{g}'(R_t^d) - \frac{\mathrm{g}(R_t^d)}{R_t^d}\right]}_{-\Omega R_t^d/L_t}\left(\frac{R_t^d}{L_t}\right) = \\ = w_t + w_t\underbrace{\left[\mathrm{N}^{R'}(X_t) - \frac{\mathrm{N}^R(X_t)}{X_t}\right]}_{(1/2)\mathrm{A}(X_t)^2/L_t}. \quad (7.15)$$

7.2.3 Market Clearing Conditions

The clearing of the commodity market in each period t demands:

$$L_t c_t^1 + L_{t-1} c_t^2 = N_t^Y \left(X_t/N_t^Y\right)^{\alpha_1}\left(K_t^d/N_t^Y\right)^{\alpha_3} - K_{t+1}. \quad (7.16)$$

The clearing of the resource spot market reads as follows:

$$R_t^d = R_t. \quad (7.17)$$

The market for capital service is cleared if the following holds:

$$K_t^d = K_t. \quad (7.18)$$

Clearing of the resource harvest market demands:

$$X_t^d = X_t. \quad (7.19)$$

Finally, the labor market clearing condition reads as follows:

$$N_t^Y + N_t^R = L_t. \quad (7.20)$$

7.3 Perfectly Competitive Markets with an Open–Access Renewable Resource

Before we embark on evaluating the efficiency of the market allocation with privately owned resource stocks, we focus in this section on the case where a renewable resource is not privately owned but is a common property resource with open access: i.e. everybody who is ready to bear the harvest costs is permitted to utilize the resource without any constraints. As a consequence of this fundamental change in the institutional framework, a market on which the resource stock is traded between subsequent generations can no longer exist. Moreover, the individual resource harvesters do not have any incentive to include in their marginal calculus the user costs of resource harvesting (which are equal to the discounted utility of future resource harvest). This disregard for part of the opportunity costs of current harvesting leads to a more intensive use of the resource (possibly overuse) than in the case where the resource is privately owned. Seen from another perspective, there is a negative stock externality caused by inappropriate property rights which induces self–interested individuals to overuse or even exhaust the renewable resource. It is thus no surprise that the market allocation is not intergenerationally efficient and that the welfare of households is less than that in the case of privately owned resources.

A very prominent example of an open-access resource, with however very low natural regeneration, is the stratosphere. Greenhouse gases are released into the stratosphere as a consequence of combustion processes with fossil fuel inputs. Since the stratosphere can be regarded as the common property of mankind, emitting greenhouse gases is equivalent to harvesting the stratosphere without paying the user costs of harvesting. This lack of property rights, combined with individual economic logic, more or less guarantees excessive use of the natural resource in that sense that collective damage incurred far exceeds the sum of individual benefits. This type of collective damage which is not taken account of by individual decision makers is called an external cost or negative externality.

Returning to our market equilibrium model with perfect competition on all existing markets, the absence of property rights can be dealt with by dropping the resource stock market both in individual optimization and in market clearing conditions. For brevity, we present only those equilibrium conditions which are affected by the lack of the resource stock market.

The first equation affected is the budget constraint of the older household. Thus, on the right hand side the revenues from sale of the resource stock vanish; leaving:

$$L_{-1}c_0^2 = r_0K_0 \cancel{+ p_0R_0}. \tag{7.21}$$

For the younger household, the modified Hotelling rule completely disappears:

$$\cancel{p_{t+1} = \frac{r_{t+1}}{\Pi - 2\Omega\left(R_t\right)} p_t}. \tag{7.22}$$

As we will soon see, most important here is the disappearance of p_{t+1}/r_{t+1} from the no–arbitrage condition determining the price of the resource harvest:

$$q_t = \cancel{\frac{p_{t+1}}{r_{t+1}}} + w_t \mathrm{N}^{R'}(X_t) = \cancel{\frac{p_{t+1}}{r_{t+1}}} + w_t \mathrm{A} X_t. \tag{7.23}$$

Moreover, the intertemporal budget constraint of the younger household is affected as follows:

$$c_t^1 + \frac{c_{t+1}^2}{r_{t+1}} + \cancel{\left(\frac{p_{t+1}}{r_{t+1}}\right)\left(g'(R_t) - \frac{g(R_t)}{R_t}\right)\left(\frac{R_t}{L_t}\right)} = \\ = w_t + w_t\left(\mathrm{N}^{R'}(X_t) - \frac{\mathrm{N}^R(X_t)}{(X_t)}\right)\left(\frac{X_t}{L_t}\right). \tag{7.24}$$

Finally, the market clearing condition for the resource stock market disappears:

$$\cancel{R_t^d = R_t}. \tag{7.25}$$

7.4 Market Equilibrium and Intergenerational Efficiency under Opposing Property Rights Regimes

This section is devoted to working out the consequences of opposing property rights regimes for intergenerational efficiency of intertemporal market equilibrium allocations. We start by showing the short-run intergenerational efficiency of perfectly competitive market equilibria when the renewable resource is privately owned.

7.4.1 Privately Owned Resource Stocks and Short-Run Intergenerational Efficiency of Market Allocation

For the sake of simplicity we set $\delta = 1$. To show that the intertemporal market equilibrium conditions imply short-run intergenerational efficiency (denoted again by $\circ$), we provisionally try the following equality settings:

$$w_t = \frac{(\phi_t^L)^\circ}{(\phi_t^Y)^\circ}, \quad p_t = \frac{(\phi_t^R)^\circ}{(\phi_t^Y)^\circ}, \quad r_{t+1} = \frac{(\phi_t^Y)^\circ (1+\theta)}{(\phi_{t+1}^Y)^\circ}, \quad \forall t. \tag{7.26}$$

We first want to show that the FOC for short-run intergenerationally efficient consumption bundles of the younger household in period t, i.e.

$$-\frac{\mathrm{d}c_{t+1}^2}{\mathrm{d}c_t^1} = \frac{c_{t+1}^2}{\beta c_t^1} = \frac{\left(\phi_t^Y\right)^\circ (1+\theta)}{\left(\phi_{t+1}^Y\right)^\circ}, \tag{6.12'}$$

holds in intertemporal market equilibrium. To see this, start from the intertemporal optimality condition (7.10) and insert the above provisional setting for r_{t+1}, noting that $\beta = 1/(1+\theta)$.

Next, we have to show that the FOC for short-run intergenerationally efficient labor input holds, i.e.

$$\left(\phi_t^Y\right)^\circ \frac{\partial \mathrm{F}}{\partial N_t^Y} \equiv (1-\alpha_1-\alpha_3)\left(\frac{X_t^d}{N_t^Y}\right)^{\alpha_1}\left(\frac{K_t^d}{N_t^Y}\right)^{\alpha_3} = \left(\phi_t^L\right)^\circ. \tag{6.13'}$$

However, this is easily shown by starting from the profit maximization condition (7.6) and inserting the provisional setting for w_t.

The FOC for the short-run intergenerationally efficient harvest volume, i.e.

$$\left(\phi_t^Y\right)^\circ \frac{\partial \mathrm{F}}{\partial X_t} \equiv \left(\phi_t^Y\right)^\circ \alpha_1 \left(\frac{X_t^d}{N_t^Y}\right)^{\alpha_1-1}\left(\frac{K_t^d}{N_t^Y}\right)^{\alpha_3} = \left(\phi_t^L\right)^\circ \mathrm{N}^{R'}(X_t) + \frac{\left(\phi_{t+1}^R\right)^\circ}{(1+\theta)} \tag{6.14'}$$

follows from the profit maximization condition (7.7), from the no-arbitrage condition (7.13) and from the provisional settings for w_t, p_t and r_{t+1}, and from the definition $\beta = 1/(1+\theta)$.

The FOC for short-run intergenerationally efficient accumulation of the natural resource stock, i.e.

$$\left(\phi_{t+1}^R\right)^\circ \left[\Pi - 2\Omega\left(R_t\right)^\circ\right] = (1+\theta)\left(\phi_t^R\right)^\circ \tag{6.15'}$$

is a consequence of the modified Hotelling rule

$$p_{t+1} = \frac{r_{t+1}}{\Pi - 2\Omega\left(R_t^d\right)} p_t, \tag{7.11}$$

market clearing condition (7.17), and of the provisional settings for r_{t+1} and p_t and the definition $\beta = 1/(1+\theta)$.

Finally, the FOC for short-run intergenerationally efficient accumulation of man-made capital, i.e.

$$\left(\phi_{t+1}^Y\right)^\circ \alpha_3 \left(\frac{X_{t+1}^d}{N_{t+1}^Y}\right)^{\alpha_1}\left(\frac{K_{t+1}^d}{N_{t+1}^Y}\right)^{\alpha_3-1} = (1+\theta)\left(\phi_t^Y\right)^\circ \tag{6.16'}$$

results if the FOC for profit maximizing capital input is written for period $t+1$ as:

$$\frac{\partial F}{\partial K_{t+1}^d} \equiv \alpha_3 \left(\frac{X_{t+1}^d}{N_{t+1}^Y}\right)^{\alpha_1}\left(\frac{K_{t+1}^d}{N_{t+1}^Y}\right)^{\alpha_3-1} = r_{t+1}, \tag{7.27}$$

and where the provisional setting for r_{t+1} and the definition $\beta = 1/(1+\theta)$ are taken into account.

To sum up, these results show that our provisional equality settings (7.26) are indeed warranted, and do in fact show that the intertemporal market equilibrium conditions imply the FOCs for short-run intergenerational efficiency. Thus, when the renewable resource is privately owned and all markets are perfectly competitive the consumption allocation associated with intertemporal market equilibrium is indeed short-run intergenerationally efficient.

7.4.2 Open Access and the Inefficiency of Market Allocation

After having shown that private ownership of the renewable resource leads to short-run intergenerational efficiency of the market allocation under perfect competition, it is not difficult to see that the market allocation is short-run intergenerational inefficient when the renewable resource can be openly accessed. The argument runs as follows: The market allocation under private ownership of the renewable resource is unique, and it is short-run intergenerationally efficient. We have shown above that the market equilibrium conditions with open access to the renewable resource differ from those under private ownership. Hence, such a market allocation cannot be short-run intergenerationally efficient. In other words: When the renewable resource can be openly accessed, the market allocation is even short-run intergenerationally inefficient. Notice, however, that the inefficiency must not be traced back to individual self-interest and market competition but rather to the lack of exclusive private property rights to the natural resource stock.

It follows almost immediately that short-run intergenerationally inefficiency leads to long-run intergenerationally inefficiency. Thus, Fig. 7.1 illustrates the long-run intergenerational inefficiency of a market allocation with open access. Here, the main reason for the inefficiency lies in the fact that with open access the resource user does not take into account the user cost of resource harvest $(\phi^R)^\bullet\,(1+\mathrm{g}(R^\bullet))$. In the upper diagram of Fig. 7.1 the wedge between the marginal value product and marginal harvest cost, as found in Fig. 6.3, is lacking, and this leads to an inefficiently high resource harvest, i.e. a level below the intersection of marginal value product and marginal harvest cost curve.

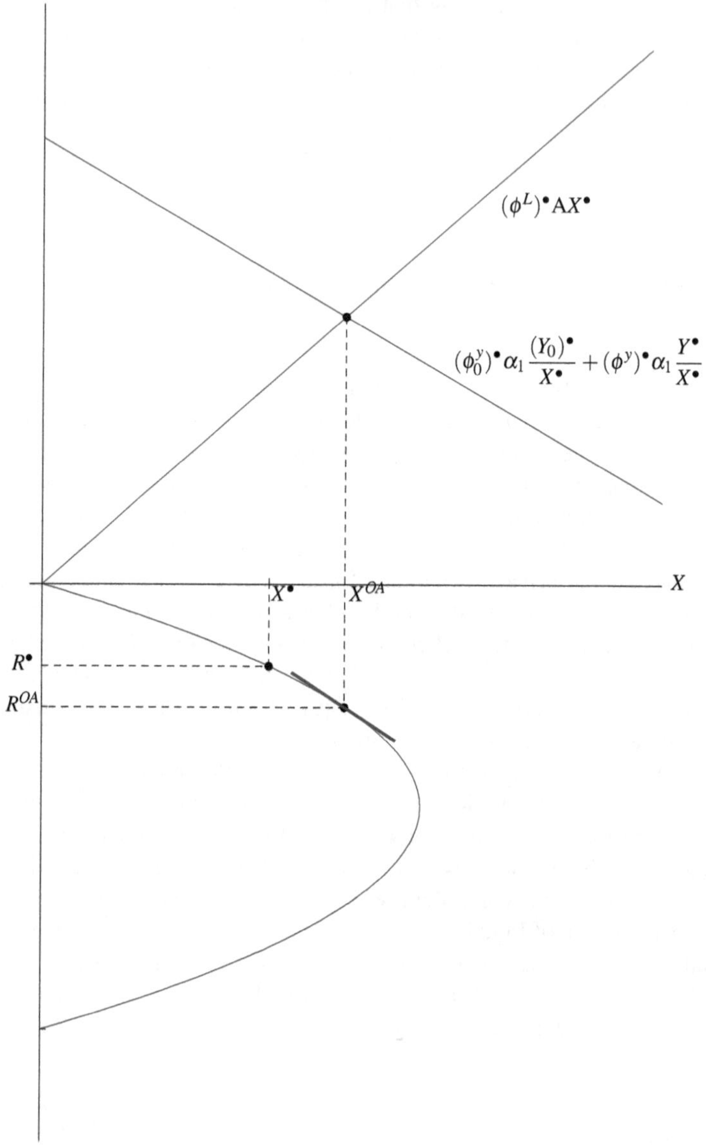

Fig. 7.1 Long-run intergenerational inefficiency of market solution for an open access resource: $R^{OA} > R^{\bullet}$

7.5 Conclusions

In this chapter we investigated intertemporal market equilibria when the stock of the renewable resource is either privately owned or can be openly accessed. Contrary to widespread thought, we were able to show that self-interested individuals are led by perfectly competitive market prices to use the renewable resource short-run intergenerationally efficient when the resource stock remains in exclusive private ownership. On the other hand, when the resource stock is openly accessed, short-run and long-run intergenerational inefficiency occur. Thus, the common presumption is true with regard to open–access resources, but not when privately owned resources are involved.

Since the costs of implementation and control of exclusive property rights in many renewable resources are prohibitive, the quest for another benchmark than intergenerational efficiency emerges. Moreover, intergenerational efficiency does not preclude significant intergenerational inequity. Hence, the question arises whether market allocations along the intertemporal equilibrium path accord with the benchmark of intergenerational equity (sustainability) or not. This question will be the subject of the next chapter.

References

Farmer K (1989) Balance sheets and budget contraints in discrete-time macro-models: a note. European Journal of Political Economy 5:533–542

Foley DK (1975) On two specifications of asset equilibrium in macroeconomic models. Journal of Political Economy 83:303–324

Part IV
Intergenerational Equity and Market Equilibrium with Scarce Renewable Resources

Chapter 8
Sustainable Economic Growth with Linear Resource Regeneration

8.1 Introduction

Until now we have focused on intergenerational efficiency as benchmark for the social evaluation of market allocation. As mentioned in the first chapter, since the 1970s the focus of resource policy turned from intergenerational efficiency to intergenerational equity, i.e. sustainability from an economic point of view. After the first oil price shock, the sustainability of economic growth (income per capita) under exhaustible (or non-renewable) became the main matter of concern. Regarding the feasibility of sustainability, we know from Chap. 2 that a constant standard of living is possible in the presence of exhaustible resources, even in the absence of technological change, provided that man-made capital and exhaustible resources are 'good' substitutes in production, and that resource owners invest sufficiently in reproducible capital so as to offset the optimally declining stock of natural resources and to achieve economic sustainability. However, this condition, known as the Hartwick (1977) rule, is derived from intertemporal equilibrium models of the infinitely-lived agent (ILA) type which ignore 'generation overlap and treat society in each period as a single generation which cares about (and also discounts) the welfare of its immediate descendants and which has complete control over the rate of resource use and the saving rate' (Mourmouras, 1991, 585).

Whether a sequence of self-interested, finite-lived and overlapping generations interacting via perfectly competitive markets will implement the ILA paths remains, as Mourmouras (1991, 585) rightly remarks, an open question. This chapter is mainly devoted to answering this question within Mourmouras' log-linear CD OLG model. In such a setting technical progress and population growth are absent, the natural resource which is necessary in production regenerates linearly, and man-made capital depreciates completely after one model period. Within such a framework, GDP growth becomes feasible because man-made capital is accumulating, despite the stock of natural resources remaining time-stationary. 'Sustainable growth' is now possible even if 'strong' ecological sustainability, in the sense of an unchanging initial stock of natural resources, is demanded.

K. Farmer and B. Bednar-Friedl, *Intertemporal Resource Economics: An Introduction to the Overlapping Generations Approach*, DOI 10.1007/978-3-642-13229-2_8,

However, sustainable growth, or more generally 'sustainable development', is not self-evident in Mourmouras' OLG model with linear regeneration. Even in the case of regenerating resources the propensity to use the resource as a production input can be too high compared to the regeneration rate of the natural resource—regardless of how well man-made capital can be substituted for the natural resource in production. Moreover, in the OLG model substitutability of capital for resources is limited by the assumption of complete depreciation of man-made capital after one period. Such circulating capital implies that in the case of exhaustible resources, where the regeneration rate is zero, a time–stationary living standard is impossible. These are results which are plainly at odds with the insights from ILA intertemporal equilibrium models (e.g. Clark, 1990).

In the next section, first the consequences of a linear regeneration function for the optimization problem of the younger household are presented. Then the intertemporal equilibrium dynamics are derived, and the linear saddle-path dynamics are analyzed. Following this, both intertemporal equilibrium and steady state growth are defined and characterized and the possibility of sustainable growth is investigated. Finally, the feasibility and necessity of sustainable growth in competitive market systems is discussed in general terms.

8.2 Individual Optimization and Market Clearing under Linear Regeneration

The main change in the model is that instead of logistic (non-linear) regeneration of the natural resource, as was assumed in the preceding chapters, the net regeneration function now reads as follows:

$$R_{t+1} = \Pi\,(R_t - X_t), \quad \Pi \geq 1.$$

This specification, which accords exactly with the Mourmouras (1991, 586) specification, cannot be obtained exactly as a special case of the logistic specification. However, a non-renewable resource does result as a special case where $\Pi = 1$.

The main impact of the change from the logistic to the linear specification of the regeneration function is felt in the FOCs of the younger household. Optimizing the utility of the younger households gives two new FOCs. The first is a Hotelling rule modified to meet the requirements of a linearly regenerating resource:

$$\frac{p_{t+1}}{p_t}\Pi = r_{t+1}, \tag{8.1}$$

which collapses to the original Hotelling (1931) rule when $\Pi = 1$.

The second FOC is a no-arbitrage condition between the harvest and resource stock price:

$$q_t = \frac{p_{t+1}}{r_{t+1}}\Pi. \tag{8.2}$$

As a result of the assumption of costless harvesting, the younger household faces the following budget constraints for the working and the retirement period:

$$c_t^1 + K_{t+1} + p_t R_t^d = w_t + q_t X_t, \tag{8.3}$$

$$c_{t+1}^2 = r_{t+1} K_{t+1} + p_{t+1} R_{t+1}. \tag{8.4}$$

The net regeneration function reads as follows:

$$R_{t+1} = \Pi R_t^d - \Pi X_t. \tag{8.5}$$

By combining the retirement period budget constraint (8.4) with the regeneration function we obtain:

$$K_{t+1} = \frac{c_{t+1}^2}{r_{t+1}} - \left(\frac{p_{t+1}}{r_{t+1}}\right) \Pi (R_t - X_t). \tag{8.6}$$

Insertion of (8.6) into the working period budget constraint, acknowledging (8.1) and (8.2) and accepting that $p_t = q_t$, $\forall t$, yields as intertemporal budget constraint:

$$c_t^1 + \frac{c_{t+1}^2}{r_{t+1}} = w_t. \tag{8.7}$$

To complete the description of the model, market clearing conditions have to be specified for all markets. Labor market clearing demands:

$$L_t = N_t^Y = 1, \quad \forall t. \tag{8.8}$$

Clearing of the resource stock market reads as follows:

$$R_t^d = R_t, \quad \forall t. \tag{8.9}$$

Resource harvest market clearing demands:

$$X_t^d = X_t, \quad \forall t. \tag{8.10}$$

Aggregate savings, defined as $s_t = w_t - c_t^1$, follow from the current period budget constraint of the younger household. This yields:

$$K_{t+1} = \sigma w_t + q_t (X_t - R_t), \tag{8.11}$$

where $\sigma \equiv \beta / (1 + \beta)$.

8.3 Derivation of Intertemporal Equilibrium Dynamics

We first derive the intertemporal equilibrium dynamics of man-made capital before deriving the dynamics of resource harvest and resource stock. Thus, on inserting $\alpha_2 Y_t = w_t$ and $\alpha_1 Y_t = q_t X_t$ into (8.11), we obtain:

$$K_{t+1} = \sigma\alpha_2 Y_t + \alpha_1 Y_t - \alpha_1 \left(\frac{Y_t}{X_t}\right) R_t. \tag{8.12}$$

Taking into account the labor market clearing condition within the aggregate production function, $Y_t = K_t^{\alpha_3} X_t^{\alpha_1}$, and inserting the result into (8.12), yields the intertemporal equilibrium dynamics, or law of motion, for man-made capital:

$$K_{t+1} = (\sigma\alpha_2 + \alpha_1)(K_t)^{\alpha_3}(X_t)^{\alpha_1} - \alpha_1 (K_t)^{\alpha_3}(X_t)^{\alpha_1 - 1} R_t. \tag{8.13}$$

To ensure non-negative stocks of man-made and natural capital, namely $K_{t+1} \geq 0$ and $R_{t+1} \geq 0$, the harvest volume over time must not become either too low or too high:

$$\left[\frac{\alpha_1}{(\sigma\alpha_2 + \alpha_1)}\right] R_t \leq X_t \leq R_t. \tag{8.14}$$

If (8.14) is strictly satisfied, it is feasible to use the no-arbitrage conditions (8.1) and (8.2), to obtain:

$$q_t = \frac{q_{t+1}}{r_{t+1}} \Pi. \tag{8.15}$$

Utilizing the profit maximizing conditions $\alpha_1 Y_t = q_t X_t$ and $\alpha_3 Y_t = r_t K_t$ in (8.15) gives:

$$\Pi \left(\frac{\frac{Y_{t+1}}{X_{t+1}}}{\frac{Y_t}{X_t}}\right) = \alpha_3 \left(\frac{Y_{t+1}}{K_{t+1}}\right). \tag{8.16}$$

The next step is to insert the CD production function into this equation which yields:

$$\Pi K_{t+1} (X_t)^{1-\alpha_1} = \alpha_3 X_{t+1} (K_t)^{\alpha_3}. \tag{8.17}$$

Inserting (8.13) into (8.17) gives the law of motion for the resource harvest:

$$X_{t+1} = \frac{(\sigma\alpha_2 + \alpha_1)}{\alpha_3} \Pi X_t - \frac{\alpha_1}{\alpha_3} \Pi R_t. \tag{8.18}$$

The law of motion for the resource stock follows immediately from the net regeneration function for a given initial value of the resource stock:

$$R_{t+1} = \Pi (R_t - X_t), \quad R_0 = \underline{R}_0 > 0. \tag{8.19}$$

In order to see whether the equilibrium dynamics is stable, we utilize the eigenvalues associated with the eigenvector of the two equations of motion. In matrix notation we thus obtain:

$$A_{t+1} = MA_t, \quad M \equiv \Pi \left(\begin{bmatrix} \frac{(\sigma\alpha_2+\alpha_1)}{\alpha_3} \\ -1 \end{bmatrix} - \begin{pmatrix} \frac{\alpha_1}{\alpha_3} \\ 1 \end{pmatrix} \right), \quad A_t \equiv \begin{pmatrix} X_t \\ R_t \end{pmatrix}.$$

As is well-known, this is a 2x2 system of homogeneous linear difference equations which can be solved recursively:

$$A_{t+1} = MA_t \quad \Rightarrow \quad A_t = M^t A_0.$$

To check the existence of a stable solution, the eigenvalues ψ and the associated eigenvectors $\tilde{A}$ of the system matrix M can be calculated by:

$$M\tilde{A} = \psi\tilde{A} \quad \Leftrightarrow \quad (M - \psi I)\tilde{A} = 0 \quad \Rightarrow \quad |M - \psi I| = 0.$$

The eigenvalues ψ for the 2x2 case can be calculated by solving the following second-order polynomial:

$$\alpha_3\psi^2 - (\alpha_3 + \sigma\alpha_2 + \alpha_1)\,\psi + \sigma\alpha_2 = 0. \tag{8.20}$$

With this so-called characteristic polynomial at hand, we are now in a position to determine the stability properties of the dynamic system (8.18)-(8.19).

Proposition 8.1 (Stability of equilibrium dynamics).

(i) If $\alpha_3 = 0$, $\psi = \dfrac{\sigma\alpha_2}{(\sigma\alpha_2 + \alpha_1)} \in (0,1)$ *and hence the equilibrium dynamics are asymptotically stable.*

(ii) If $\alpha_3 > 0$, $_2\psi_1 = \dfrac{\alpha_3 + \sigma\alpha_2 + \alpha_1 \pm \sqrt{(\alpha_3 - \sigma\alpha_2)^2 + \alpha_1^2 + 2\alpha_1(\alpha_3 + \sigma\alpha_2)}}{(2\alpha_3)}$ *with* $\psi_1 > 1$, $0 < \psi_2 < \frac{\sigma\alpha_2}{(\sigma\alpha_2+\alpha_1)}$ *and hence the equilibrium dynamics are saddle-path stable.*

Proof. Mourmouras (1991, Appendix). □

This proposition tells us that in case (i) in which man-made capital is not productive ($\alpha_3 = 0$) the equilibrium dynamics of the resource stock and its harvest are asymptotically stable because the single eigenvalue is larger than zero and less than unity (The equilibrium dynamics is only one-dimensional). In case (ii) we encounter a truly two-dimensional equilibrium dynamics with two distinct, real eigenvalues $\psi_1 > 1$ and $0 < \psi_2 < 1$. Since the first eigenvalue is larger than one, while the second eigenvalue lies between zero and one, we face in case (ii) saddle-path dynamics which is only 'knife-edge' stable. Saddle-path stability implies that one of the two dynamical variables must be a so-called 'jump' variable for which no exogenous initial value can be assumed. The jump variable in our OLG model is represented by the resource harvest X_t or the associated resource price p_t (q_t).[1]

The eigenvalues are not only useful in determining the stability properties of a dynamic system. They can also be used to characterize the equilibrium dynamics, (8.13), and (8.18)-(8.19).

Proposition 8.2. *Suppose that* $\Pi \geq 1, \alpha_3 > 0$ *and* $\psi = \psi_2$. *Then we get as intertemporal paths of the resource stock, its harvest, and man-made capital:*

$$R_{t+1} = \Pi \psi R_t,\ R_0 = \underline{R}_0, \tag{8.21a}$$

$$X_t = (1-\psi) R_t, \tag{8.21b}$$

$$K_{t+1} = \left[\sigma\alpha_2 - \frac{\psi}{1-\psi}\alpha_1\right] (X_t)^{\alpha_1} (K_t)^{\alpha_3}, \quad K_0 = \underline{K}_0 > 0. \tag{8.21c}$$

Some remarks are now needed in order to better understand the meaning of Prop. 8.2. First, while both stock variables require an initial value for the dynamic system to be fully specified, the initial value of resource harvest is determined implicitly by (8.21b) in which the dynamics of the eigenvector, associated with the second eigenvalue, is depicted. Second, Prop. 8.2 provides an alternative but equivalent representation of the original equilibrium dynamics with the advantage that the qualitative information, provided by Prop. 8.1, can be utilized for analyzing the properties of the equilibrium dynamics more deeply. This is the subject of the next section.

8.4 Intertemporal-Equilibrium and Steady-State Growth Rates

In this section we want to define and characterize the main model variables along an intertemporal equilibrium path and also in a steady state. While we do not assume exogenous population and/or factor productivity growth endogenous steady state growth results on account of natural growth. We start with the definition of the intertemporal and the steady-state growth rate rate of any variable x_t in our log-linear CD OLG model. The intertemporal growth rate for a variable x is defined by $g_t^x \equiv \ln x_{t+1} - \ln x_t$, while the steady state growth rate is given by $g^x \equiv \lim_{t\to\infty} g_t^x$, or equivalently, $g_{t+1}^x = g_t^x = g^x$. With these definitions at hand, we are ready to characterize the intertemporal equilibrium growth rates of the main economic variables.

Corollary 8.1 (Intertemporal equilibrium growth rates).

$$g_t^R = g_t^X = \ln(\Pi\psi), \tag{8.22}$$

$$g_{t+1}^K = \alpha_1 \ln(\Pi\psi) + \alpha_3 g_t^K, \tag{8.23}$$

$$g_t^Y = \alpha_3 g_t^K + \alpha_1 g_t^X, \tag{8.24}$$

[1] For a more extensive discussion of the role of jump variables in saddle-path systems emanating from Diamond-type OLG models, see Farmer (2006).

$$g_t^w = g_t^Y, \tag{8.25}$$

$$g_t^r = g_t^Y - g_t^K, \tag{8.26}$$

$$g_t^q = g_t^Y - g_t^X. \tag{8.27}$$

Proof. The proof consists in applying the definition of the growth rates to the equilibrium dynamics in Prop. 8.2. Since this is straightforward, we provide only one case. To show that $g_{t+1}^K = \alpha_1 \ln(\Pi\psi) + \alpha_3 g_t^K$ is indeed the case, take the natural logarithm of both sides of (8.21c). This yields for $t+1$:

$$\ln K_{t+1} = \ln\left[\sigma\alpha_2 - \frac{\psi}{1-\psi}\alpha_1\right] + \alpha_1 \ln X_t + \alpha_3 \ln K_t,$$

and a similar expression can be derived for t. Subtracting the latter from the former equation, we get:

$$\ln K_{t+1} - \ln K_t = \alpha_1(\ln X_t - \ln X_{t-1}) + \alpha_3(\ln K_t - \ln K_{t-1}).$$

□

Corollary 8.2 (Existence, uniqueness and stability of steady state growth). *A non-trivial steady state growth rate of man-made capital exists, is unique and globally stable:*

$$g^K = \frac{\alpha_1}{\alpha_2 + \alpha_1} g^R. \tag{8.28}$$

At the unique steady state, man-made capital and output grow at a similar rate:

$$g^Y = g^K. \tag{8.29}$$

Furthermore, resource harvest price grows at

$$g^q = -\frac{\alpha_2}{\alpha_2 + \alpha_1} g^R, \tag{8.30}$$

and the growth rate of the interest factor on man-made capital is zero:

$$g^r = 0. \tag{8.31}$$

Proof. Here we merely show that $g^Y = g^K$. From $g_t^Y = \alpha_1 g_t^X + \alpha_3 g_t^K$ and knowing that $g^K = \frac{\alpha_1}{\alpha_2+\alpha_1} g^R$ we obtain

$$g^Y = \alpha_1 g^X + \alpha_3\alpha_1/(\alpha_1+\alpha_2) g^R = \alpha_1/(\alpha_1+\alpha_2) g^X = g^K, \tag{8.32}$$

since $\alpha_1 + \alpha_2 + \alpha_3 = 1$. □

The results of both corollaries have interesting interpretations. First, the stock and the harvest of the natural resource grow by the same rate $\ln(\Pi\psi)$. Clearly, the growth rate is larger than zero if $\ln(\Pi\psi) > 1$. While this growth rate is constant over time, the growth rate of man-made capital (accumulation rate) changes along the intertemporal equilibrium path. Since $\alpha_3 < 1$, it converges asymptotically towards $g^K = \alpha_1/(\alpha_2 + \alpha_1)g^R$. Thus, the two stocks grow differently in steady state. We also see that $g^r = 0$ and $g^q = -(\alpha_2/(\alpha_1 + \alpha_2))g^R$. Hence, the real interest rate is time constant and the growth rate of the harvest price is negative (positive) when the growth rate of the resource stock and resource harvest is positive (negative).

Moreover, the growth properties of the model depend on the product of the regeneration factor Π, and the equilibrium propensity to invest in resources ψ. If $\psi\Pi < 1$, man-made and natural capital decline geometrically. This is in fact the case if the resources are exhaustible and thus $\Pi = 1$. In this steady state, resource prices grow at a constant rate while the marginal product of capital and the rate of interest remain constant. Constancy of the marginal product of K is possible with declining levels of K because X declines even faster. Here we encounter the case of an exponentially declining world economy in spite of man-made capital being a good substitute for natural resources.

8.5 A Sustainable Economic Growth Path

The case $\psi\Pi = 1$ has particularly 'appealing properties because it provides for natural resources to be distributed in an egalitarian fashion across generations *and* simultaneously allows for improvements in standards of living to take place through capital accumulation' (Mourmouras, 1991, 589). It is apparent that this appealing case is not feasible in the case of exhaustible resources, since then on account of $\psi < 1$ the case $\psi\Pi = 1$ is excluded.

We focus now on the sustainable development path where $\psi\Pi = 1$, presupposing Π sufficiently larger than one. This leads to the following proposition on sustainable economic growth or development which requires certain combinations of parameter values for the subjective time preference rate, output elasticities, and the natural regeneration rate such as to ensure that $\psi\Pi = 1$.

Proposition 8.3. *(Sustainable economic growth) If the 'deep' parameters of the OLG economy with regenerating resources are such that $\Pi\psi = 1$, then economic growth is ecologically sustainable.*

Proof. On inserting $\Pi\psi = 1$ into (8.21a), we obtain: $R_t = \underline{R}_0, \forall t$. Acknowledging $\psi = \Pi^{-1}$ in (8.21b) implies for sustainable resource harvest:

$$X_t = X = \left[\frac{\pi}{\Pi}\right] \underline{R}_0. \tag{8.33}$$

Proceeding similarly for (8.21c) yields for sustainable accumulation of man-made capital:

$$K_{t+1} = \left[\sigma\alpha_2 - \frac{\alpha_1}{\pi}\right] \left(\frac{\pi}{\Pi}\right)^{\alpha_1} (\underline{R}_0)^{\alpha_1} (K_t)^{\alpha_3}, \tag{8.34}$$

which reduces at the steady state to:

$$K = \left\{ \left[\sigma\alpha_2 - \frac{\alpha_1}{\pi} \right] \left(\frac{\pi}{\Pi} \right)^{\alpha_1} (\underline{R}_0)^{\alpha_1} \right\}^{\frac{1}{1-\alpha_3}}. \tag{8.35}$$

If moreover $\underline{K}_0 < K$, then the growth rate of man-made capital is positive ($g_t^K > 0$) and follows the following geometric series:

$$g_{t+1}^K = \alpha_3 g_t^K. \tag{8.36}$$

□

Remark 8.1. The sustainably growing economy converges to a stationary state in which $g^K = g^Y = 0 = g^w$ holds.

Along the sustainable growth path towards the stationary state the real interest rate is higher and the resource price is lower than in the stationary state. While the resource stock remains constant over time at the initial value, resource prices rise over time, although at a declining rate, and the real interest rate falls. Not surprisingly, in the stationary state both the resource prices and the real interest rate are stationary, and the real interest rate is equal to π, called by Samuelson (1958) the biological interest rate. Notice also that in contrast to the growth dynamics of the Diamond OLG model without natural resources (Diamond, 1965) the stationary state emanating here from sustainable growth depends on initial conditions, in particular the economy's initial endowment with natural resources (see (8.35)).

8.6 Feasibility of Sustainable Growth and Sustainability Policy in Market Systems

The previous section has shown that even in an unhampered, fully competitive market system with linearly renewable resources, sustainable growth (development) is feasible. Economic growth and ecological sustainability are thus not fundamentally inconsistent. Even if we disregard for the moment the restrictiveness of the linearity assumption the question remains as to how realistic are the constellations of deep model parameters, like the subjective rate of time preference, the output elasticities of production factors, and the natural regeneration rate, that imply $\Pi\psi = 1$?

Without going into an extensive empirical investigation we would still like to check the plausibility of the deep parameter values leading to sustainable growth constellations. We now distinguish two main cases.

First, consider the rather simple case of $\alpha_3 = 0$ and assume that $\alpha_2 = 2/3$ and therefore $\alpha_1 = 1/3$. Considering $\Pi\psi = 1$ in the determination of the eigenvalue ψ, according to case (i) of Prop. 8.1, yields:

$$\alpha_2(\Pi - 1)\beta = (1+\beta)\alpha_1.$$

On supposing furthermore that $\Pi = 2$, it follows that $\beta/(1+\beta) = 1/2$ which is equivalent to $\theta = 0$. Hence, with a labor share of two thirds and a net regeneration rate of 1, i.e. the unused resource doubles every generation (a period of 30 years), the subjective time preference rate, $\theta = \beta^{-1} - 1$, needs to be zero in order to allow for a sustainable growth path. If the subjective rate of time preference is larger than zero, then the natural regeneration rate has to be higher too.

Second, if $\alpha_3 > 0$, the condition $\Pi\psi = 1$ becomes non-linear. Experimentation with alternative values of the time preference rate and the net regeneration rate, for given parameter values of production elasticities, shows that greater impatience (i.e. a higher time preference factor) demands higher net regeneration rates to ensure sustainable growth. However, notice that with a rising production elasticity of capital the regeneration rate can decline while still enabling a positive time preference rate. This is due to the fact that a higher production elasticity of capital implies that the natural resource becomes, relative to man-made capital, less important in generating output.

Eventually, the question arises how the people and their political representatives will respond when faced with a non-sustainable growth path in a perfectly competitive market economy? This question is extremely relevant since sustainable growth is only knife-edge stable. As a consequence, even a slight deviation will lead to an implosion of the economy in finite time. If the citizens and their political representatives can foresee such a destructive development they will probably opt for political restrictions of resource utilization by individuals (firms included). Since, however, political intervention within a free market system of decentralized decision making causes efficiency losses and hinders innovation dynamics, the sustainability objective needs to be pursued with considerable caution and skill.

In a situation where neither the people nor the politicians are concerned with ecological sustainability, economic contraction remains a real possibility in OLG economies in contrast to the expected outcome in intertemporal equilibrium models with infinitely lived dynasties.

8.7 Conclusions

Under linear regeneration, ecological sustainability and economic growth are not fundamentally inconsistent in a perfectly competitive market economy. Such a sustainable growth path is however only knife-edge stable. The slightest changes in deep parameters can trigger off an implosion of the economic system in finite time.

Moreover, in spite of man-made capital being a good substitute for natural resources, generation overlap, finite lives of generations, and complete depreciation of real capital after a single generation prevent constant standards of living when the natural resource is exhaustible. This result contrasts starkly with that in ILA intertemporal equilibrium models.

Although a 'sustainable development' path in an unhampered market economy is achievable in principle, it is unrealistic to presume that such a development will be implemented automatically via self-interest and market competition. On the other hand, ecologically motivated political intervention in individual decision making may also produce counterproductive results as shown for example by Ono (2002). Sustainable development is thus best seen as an ethical point of reference both for individuals and for politicians—a benchmark in constant need of analysis and review.

In this chapter we assumed linear regeneration of the renewable resource. A subsequent question is therefore whether the main results change when resource regenerate non–linearly. This will be investigated in the next chapter.

References

Clark CW (1990) Mathematical Bioeconomics: The Optimal Management of Renewable Resources, 2nd edn. Wiley, NY

Diamond P (1965) National debt in a neoclassical growth model. American Economic Review 55:1126–1150

Farmer K (2006) Reducing public debt under dynamic efficiency: Transitional dynamics in Diamond's OLG model. Atlantic Economic Journal 34:195–208

Hartwick J (1977) Intergenerational equity and the investing of rents from exhaustible resources. American Economic Review 66:792–794

Hotelling H (1931) The economics of exhaustible resources. Journal of Political Economy 39:137–175

Mourmouras A (1991) Competitive equilibria and sustainable growth in a life–cycle model with natural resources. Scandinavian Journal of Economics 93(4):585–591

Ono T (2002) The effects of emission permits on growth and the environment. Environmental and Resource Economics 21:75–87

Samuelson PA (1958) An exact consumption-loan model of interest with or without the social contrivance of money. Journal of Public Economics 66:467–482

Chapter 9
Steady-State Sustainability under Logistically Regenerating Resources

9.1 Introduction

While after the first oil price shock the sustainability of economic growth under *exhaustible* resources was the main matter of concern, the Brundtland report (WCED, 1987) brought the problem of the sustainability of *renewable* resources to the attention of the world community. Given the prominence of the sustainability criterion for renewable resources, it is surprising that so little has been done to investigate the question of whether an unhampered, competitive market economy which utilizes a renewable resource both as productive input and as an asset is at all capable of generating intertemporal equilibrium paths with either an egalitarian distribution of utilities (consumption) of subsequent generations or an egalitarian distribution of resource stocks across generations (ecological sustainability).

As we know from the previous chapter, Mourmouras (1991) has investigated the feasibility of the latter ecological sustainability in a Diamond (1965) type overlapping generations' market equilibrium model in which a renewable resource is essential for production, technological progress is absent, and man-made capital circulates. Given specific values for the time-preference parameter, for the output elasticities of the production factors (labor, man-made capital, resource harvest), and for the growth factor of the renewable resource, an intertemporal equilibrium path exists along which man-made capital accumulates, the GDP growth is positive, and the distribution of resource stock across generations is such that 'each generation [is ready] to reserve for the future the stock of the natural resources it has inherited from the past' (Mourmouras, 1991, 586).

To show the feasibility of such sustainable growth in a perfectly competitive market economy over time, Mourmouras presupposes a linear regeneration function which implies that the (exponential) growth rate of the resource does not depend on the resource stock (see Chap. 8). This specification simplifies the analysis of the intertemporal market equilibrium dynamics, but it also provokes the question of whether man-made capital accumulation and time-stationary stocks of the renewable resource are also to be expected under a non-linear regeneration function (e.g.

K. Farmer and B. Bednar-Friedl, *Intertemporal Resource Economics: An Introduction to the Overlapping Generations Approach*, DOI 10.1007/978-3-642-13229-2_9,

a logistic one as in Chaps. 6–7) which captures the essentials of most renewable resources better than Mourmouras' linear case which is relevant for expendables (e.g., the use of hydropower). If—as may be easily suggested—Mourmouras' nice sustainable-growth result does not hold in the more general case of logistic regeneration we have to face the question which sustainability result if any can be expected in this case.

To anticipate the answer to this question it is only stationarity state sustainability we will obtain under logistic regeneration. However, contrary to first thought, neither the existence nor the dynamic stability of a steady-state solution are guaranteed without restrictive assumptions with respect to the parameters of the utility, production and regeneration function as Farmer (2000) has first shown. Thus, this chapter is devoted to proving rigorously both the existence and dynamic stability of steady states which are both economically and ecologically sustainable almost by definition. Furthermore, we will show the infeasibility of sustainable growth under logistic regeneration.

The chapter is organized as follows: In the next section the log-linear CD OLG model of the Chap. 7 is rapidly reviewed. Then, the two-dimensional equilibrium dynamics of the resource stock and the resource harvest is explicitly derived, and the sufficient conditions for the existence of non-trivial steady states (fixed points of the equilibrium dynamics) are investigated, along with the local dynamic (in-)stability of the trivial and non-trivial steady state. With the aid of the saddle-path of the equilibrium dynamics we finally show the impossibility of ecological sustainability under logistic regeneration.

9.2 The Log-Linear CD OLG Model with Logistic Regeneration

The basic characteristics of the OLG model of this chapter, which was already used in Chap. 7, has two important differences to Mourmouras' (1991) specification used in the previous chapter: the first concerns the already mentioned nonlinearity of the regeneration function, the second the ownership of the resource stock. While Mourmouras (1991) did not explicitly define the structure and the transaction possibilities with respect to the property rights for the resource, within the present model the resource is privately owned by the households and traded between the older and the younger generation in every market period. In contrast to Chap. 7 there are no harvest costs in this chapter, and hence $A = 0$.

As in Mourmouras (1991), the stock of man-made capital depreciates completely after one period. On the other hand, the natural-resource stock is assumed to be durable. The reallocation of the already existent resource stock between traders can be distinguished from the coordination of resource flows harvested during the market period. To provide a glimpse of this stock-flow distinction in a discrete-time, intertemporal equilibrium model (Farmer, 1989), the already existing stocks (of the natural resource) are traded between (old and young) households in 'beginning-of-period' ('spot') markets (Foley, 1975), the flow of the resource harvest is traded between the young household and the producer in 'end-of-period' ('one-period forward') (Foley, 1975) markets.[1] Transaction costs and other barriers to trade are absent in both markets.

This setup allows us to describe the budget constraints of the agents. Let us start with the older generation (superscript 2) in market period $t = 0$. At the beginning of this market period, the older generation uses the proceeds from its resource stock R_t supplied at the spot price p_t (all prices are calculated in terms of the single commodity produced) and its rental income $r_t K_t$, defined by the rental rate r_t times the supply of the stock of man-made capital K_t, to finance its retirement consumption c_t^2. Notice that we assume again both no population growth and $L_{-1} = 1$, hence c_t^2 denotes both individual and aggregate consumption of the older household. Similar is true for younger households.

The younger generation (superscript 1) receives the wage income w_t in exchange for its labor supply of 1 unit, and also the revenues from the supply X_t of the resource harvest at the one-period forward price q_t. These revenues are used to purchase the renewable resource-stock quantity R_t at the spot price p_t, the consumption 'flow' c_t^1, and the (gross) investment of man-made capital K_{t+1}.

The unused resource stock R_{t+1} is sold at the beginning of period $t+1$ at the spot price p_{t+1} expected by the currently younger households under perfect foresight. The revenues from stock sales and the rental income from the man-made capital stock $r_{t+1}K_{t+1}$ are used to finance the retirement consumption c_{t+1}^2 of the older generation in $t+1$.

The typical firm maximizes the profit in each period $t = 0, \ldots$. The optimization problem takes the following form:

$$\max \left(X_t^d\right)^{\alpha_1} \left(N_t^Y\right)^{\alpha_2} \left(K_t^d\right)^{\alpha_3} - w_t N_t^Y - q_t X_t^d - r_t K_t^d$$

$$\text{subject to: } N_t^Y \geq 0,\ X_t^d \geq 0,\ K_t^d \geq 0,\ \alpha_1 + \alpha_2 + \alpha_3 = 1.$$

The production elasticities of labor, resources, and man-made capital are denoted by α_1, α_2 and α_3 respectively, whereby $0 < \alpha_i < 1$, $i = 1, 2, 3$. N_t^Y is the labor demand, X_t^d is the demand for the resource harvest, and K_t^d is the demand for capital services of the firm in period t.

The typical younger household solves the following intertemporal utility maximization problem:

$$\max \rightarrow \ln c_t^1 + \beta \ln c_{t+1}^2$$

subject to:

[1] We would like to alert the reader that this beginning-of-period formulation of the resource stock market is unconventional in the literature on discrete time intertemporal resource economic models (e.g. Olson and Knapp, 1997; Koskela et al, 2002). However, as shown in Farmer (2000), under the end-of-period specification of the resource-stock equilibrium the optimization conditions of young households change only slightly; merely a formally more complex resource harvesting dynamics results without altering the main substantial arguments of this chapter.

$$p_t R_t + c_t^1 + K_{t+1} = q_t X_t + w_t, \tag{9.1a}$$

$$c_{t+1}^2 = r_{t+1} K_{t+1} + p_{t+1} R_{t+1}, \tag{9.1b}$$

$$R_{t+1} = R_t + g(R_t) - X_t, \tag{9.1c}$$

$$c_t^1 \geq 0,\ R_t \geq 0,\ X_t \geq 0,\ c_{t+1}^2 \geq 0,\ K_{t+1} \geq 0,\ R_{t+1} \geq 0. \tag{9.1d}$$

$0 < \beta < 1$ denotes the subjective time-discount factor, $g(\cdot)$ is the regeneration function with $g'' < 0$, $g(0) = 0$, and $g(R_{max}) = 0$ for $R \geq R_{max} > 0$. As a parametric example, we will again use the logistic function $g(\cdot) \equiv (\Pi - 1) R_t - \Omega (R_t)^2$ with $\Pi > 1$ as the constant-growth factor and $\Omega \ll 1$ as the growth-retarding factor of resource accumulation. (9.1a) is the current, and (9.1b) the future budget constraint of the younger household, (9.1c) represents the net regeneration function.

At an interior solution of the household's optimization problem the following relationships hold:

$$\frac{c_{t+1}^2}{r_{t+1}} = \beta c_t^1, \tag{9.2}$$

$$q_t = \frac{p_{t+1}}{r_{t+1}}, \tag{9.3}$$

$$\left(1 + g'\left(R_t^{\mathrm{d}}\right)\right) \frac{p_{t+1}}{p_t} = r_{t+1}. \tag{9.4}$$

In (9.2) discounted marginal utilities are equalized; (9.3) represents the no-arbitrage condition when deciding to harvest the resource or not, and (9.4) is the Hotelling (1931) rule in our model of renewable natural-resource allocation over time.

To derive the intertemporal budget constraint, solve (9.1b) K_{t+1} and insert (9.1c) for R_{t+1} to obtain

$$K_{t+1} = \frac{c_{t+1}^2}{r_{t+1}} - \frac{p_{t+1}}{r_{t+1}} \left[R_t + g(R_t) - X_t\right].$$

Then, insert the expression for K_{t+1} into (9.1a). This yields:

$$c_t^1 + \frac{c_{t+1}^2}{r_{t+1}} + R_t \left\{ p_t - \frac{p_{t+1}}{r_{t+1}} \left[1 + \frac{g(R_t)}{R_t}\right] \right\} + X_t \left[\frac{p_{t+1}}{r_{t+1}} - q_t\right] = w_t.$$

Consideration of (9.3) and (9.4) leads to the intertemporal budget constraint:

$$c_t^1 + \frac{c_{t+1}^2}{r_{t+1}} = w_t + q_t \left[\frac{g\left(R_t^{\mathrm{d}}\right)}{R_t^{\mathrm{d}}} - g'\left(R_t^{\mathrm{d}}\right)\right] R_t^{\mathrm{d}}. \tag{9.5}$$

Constraint (9.5) needs some additional comments. First, it claims that the young household's present discounted value of consumption does not just equal current wage income (as in the traditional Diamond, 1965, or in the Mourmouras, 1991, OLG model) but transgresses it by the real present value of the rent (=difference between average and marginal productivity) of the renewable resource. This means that the endowment of the young household consists not only of its labor power but also of the resource it owns. The resource rent is positive because the natural production function is assumed to be strictly concave. The resource rent would disappear if the regeneration function were linear as in the previous chapter.

The second comment in regard to (9.5) is that the resource rent accrues to the young and not to the old generation because the former has purchased the resource stock in the beginning-of-period resource stock market and is currently entitled to receive the rent which the strictly concave 'technology' of the renewable resource is generating.[2]

From (9.5) and (9.2) one derives for the consumption level of the current younger generation:

$$c_t^1 = \gamma \left\{ w_t + q_t \left[\frac{g(R_t)}{R_t} - g'(R_t) \right] R_t \right\}, \; \gamma \equiv \frac{1}{1+\beta}, \tag{9.6}$$

while the consumption level of the current older generation is determined by:

$$c_t^2 = r_t K_t + q_t \left[1 + g'(R_t) \right] R_t. \tag{9.7}$$

Profit maximization implies:

$$q_t X_t^d = \alpha_1 Y_t, \qquad w_t N_t^Y = \alpha_2 Y_t, \qquad r_t K_t^d = \alpha_3 Y_t. \tag{9.8}$$

Additionally, intertemporal equilibrium requires the clearing of the resource stock market, the clearing of the markets of capital and labor services, the clearing of the resource harvest market, and the clearing of the output market for all t:

$$R_t^d = R_t, \; \forall t, \tag{9.9}$$

$$K_t^d = K_t, \; \forall t, \tag{9.10}$$

$$N_t^Y = 1, \; \forall t, \tag{9.11}$$

$$X_t^d = X_t, \; \forall t, \tag{9.12}$$

$$\left(X_t^d \right)^{\alpha_1} \left(N_t^Y \right)^{\alpha_2} \left(K_t^d \right)^{\alpha_3} = c_t^1 + c_t^2 + K_{t+1}, \; \forall t. \tag{9.13}$$

Due to Walras law, (9.13) is redundant.

[2] In an economy in which each generation consists of many (a continuum) of individuals, intragenerational externalities among individual marginal resource productivities are likely to arise. Assuming 'rivalrous (depletable) externalities' (Baumol and Oates, 1988, 20) as well as well-defined and enforceable property rights specified over the externalities, our analysis remains true even if the notion of a representative agent is not taken literally.

9.3 Intertemporal Equilibrium Dynamics

Combining the above stated individual optimization (9.5)-(9.8) and market clearing (9.9)-(9.13) conditions, the intertemporal equilibrium dynamics of resource harvest and the stocks of man-made capital and the natural resource can then be derived.

Adapting (9.6), (9.7), and (9.13), in order to take (9.10)-(9.11) into account and inserting the revised (9.6) and (9.7) in the revised (9.13) gives:

$$K_{t+1} = Y_t - \gamma [w_t + q_t \phi(R_t) R_t] - r_t K_t - q_t [1 + g'(R_t)] R_t, \qquad (9.14)$$

with $\phi(R_t) \equiv g(R_t)/R_t - g'(R_t)$. Equations (9.9), (9.10), and (9.12) in (9.8) and revised (9.8) in (9.14) yield:

$$K_{t+1} = Y_t - \gamma\alpha_2 Y_t - \gamma\alpha_1 \frac{Y_t}{X_t} \phi(R_t) R_t - \alpha_3 Y_t - \alpha_1 \frac{Y_t}{X_t} [1 + g'(R_t)] R_t. \qquad (9.15)$$

Adjusting (9.4) to take account of (9.8) yields:

$$K_{t+1} = \alpha_3 \left[\frac{X_{t+1}}{1 + g'(R_{t+1})} \right] \left(\frac{Y_t}{X_t} \right). \qquad (9.16)$$

Since $(R_{t+1}, K_{t+1}) \geq 0, \forall t$ it is required that

$$\{\gamma\phi(R_t) + [1 + g'(R_t)]\} \frac{\alpha_1 R_t}{\alpha_1 + \alpha_2 \sigma} \leq X_t \leq R_t + g(R_t), \qquad (9.17)$$

for such R_t that $1 + g'(R_t) \geq 0, \forall t$.

Assuming that (9.17) holds, the right-hand sides of (9.15) and (9.16) may be equated. Rearranging yields the law of motion of resource harvest:

$$X_{t+1} = \left[1 + \frac{g'(R_{t+1})}{\alpha_3} \right] \{(\alpha_1 + \alpha_2 \sigma) X_t - \\ - \alpha_1 \gamma \phi(R_t) R_t - \alpha_1 [1 + g'(R_t)] R_t\}, \qquad (9.18)$$

with $\sigma \equiv 1 - \gamma$.

The second equation of motion is identical with the regeneration function:

$$R_{t+1} = R_t + g(R_t) - X_t. \qquad (9.19)$$

Inserting the production function in (9.16) gives:

$$K_{t+1} = \alpha_3 \left[\frac{X_{t+1}}{1 + g'(R_{t+1})} \right] (X_t)^{\alpha_1 - 1} (K_t)^{\alpha_3}. \qquad (9.20)$$

Equations (9.18) and (9.19) represent the fundamental laws of motion of renewable resource utilization in our intertemporal equilibrium model. Since the man-made capital stock does not appear in (9.18) and (9.19), it is sufficient to study the properties of the two-dimensional nonlinear difference-equation system (9.18) and (9.19). The dynamics of man-made capital accumulation then simply follows from (9.20).

The paths where the resource stock and resource harvesting are weakly increasing over time $R_{t+1} \geq R_t$ and $X_{t+1} \geq X_t$ follow from (9.19)

$$R_{t+1} \geq R_t \Longleftrightarrow g(R_t) \geq X_t, \tag{9.21}$$

and from (9.18)

$$X_{t+1} \geq X_t \Longleftrightarrow \left[1 + g'(R_{t+1})\right] \{(\alpha_1 + \alpha_2 \sigma) X_t - \\ - \alpha_1 \gamma \phi (R_t) R_t - \alpha_1 \left[1 + g'(R_t)\right] R_t\} \geq \alpha_3 X_t. \tag{9.22}$$

9.4 The Existence of (Non-Trivial) Steady States

Steady states (fixed-point solutions) of the equilibrium dynamics (9.18) and (9.19) are defined by $X_{t+l} = X_t = X$ and $R_{t+1} = R_t = R$ for $t \to \infty$. A glance at (9.18) and (9.19) evaluated at the so-defined steady state reveals that $(X, R) = (0, 0)$ is an admissible but trivial steady state since by assumption $g(0) = 0$. But there are also non-trivial steady states which are characterized by the following equations:

$$X = g(R), \tag{9.23}$$

$$\frac{\alpha_3 X}{\{(\alpha_1 + \alpha_2 \sigma) X - \alpha_1 \gamma \phi (R) R - \alpha_1 [1 + g'(R)] R\}} = \left[1 + g'(R)\right]. \tag{9.24}$$

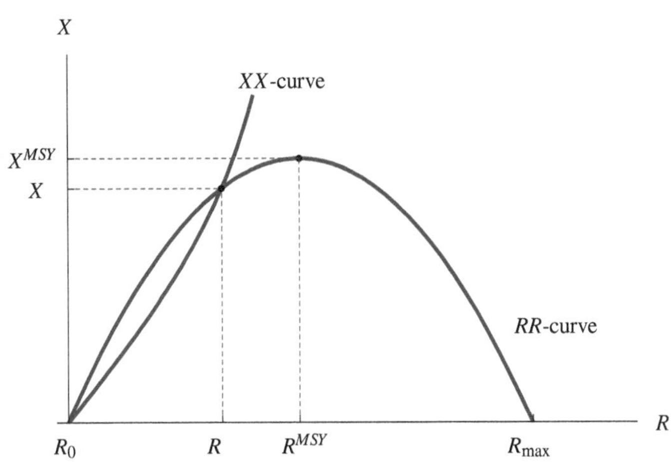

Fig. 9.1 *RR*- and *XX*-curves with steady state $R < R^{MSY}$ and $X < X^{MSY}$

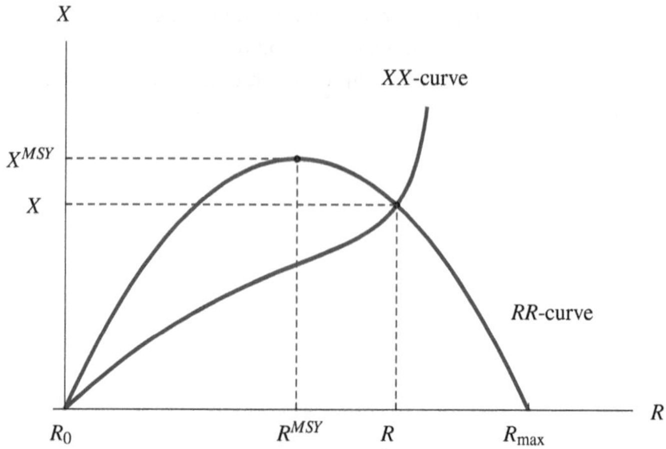

Fig. 9.2 *RR*- and *XX*-curves with steady state $R > R^{MSY}$ and $X < X^{MSY}$

These define two curves in the *RX*-space for which the resource stock and resource harvesting remain time-stationary. Presupposing the above mentioned parametric example of the natural-growth function and given plausible parameter values for the utility, production, and natural-growth function, a numerically specified example of (9.23) is depicted as *RR*-curve and an example of (9.24) is graphed as *XX*-curve in Figs. 9.1 and 9.2. Noting (9.21) and (9.22), it is easily seen that the resource stock accumulates (decreases) below (above) the *RR*-curve, while the harvesting of the resource stock increases (decreases) over time above (below) the *XX*-curve.

To be able to answer the existence question more generally, mathematical analysis is necessary. Inspecting (9.24), a strictly positive resource harvesting requires $1+g'(\mathrm{R})>0$. Given that we can write (9.24) equivalently as

$$X=\frac{\alpha_1\left[1+g'(R)\right]\left\{\gamma\phi(R)+\left[1+g'(R)\right]\right\}R}{\left[1+g'(R)\right](\alpha_1+\alpha_2\sigma)-\alpha_3}\equiv f(R)\,. \tag{9.25}$$

A non-trivial steady state exists if $0<R<R_{max}$ such that the right-hand sides of (9.23) and of (9.25) are equal and X resulting from (9.25) fulfills (9.17).

Differentiation of (9.23) and (9.25) yields for the slope of the *RR*- and *XX*-curve respectively:

$$\left.\frac{\mathrm{d}X}{\mathrm{d}R}\right|_{R_{t+1}=R_t}=g'(R)\,, \tag{9.26}$$

$$\left.\frac{\mathrm{d}X}{\mathrm{d}R}\right|_{X_{t+1}=X_t}=\frac{\alpha_1\left[1+g'(R)\right]\left\{1+g'(R)\sigma g''(R)R\right\}-\alpha_3\dfrac{g''(R)g(R)}{1+g'(R)}}{(\alpha_1+\alpha_2\sigma)\left[1+g'(R)\right]-\alpha_3}\,. \tag{9.27}$$

The slope of the RR-curve in (9.26) may be positive or negative depending on whether the steady state resource stock is less than or greater than the resource stock giving the maximum sustained yield. To determine the sign of the derivative in (9.27), conditions for the existence of a non-trivial (strictly positive) steady state are required.

As in the standard Diamond (1965) OLG model without natural resources, for which Galor and Ryder (1989, 369-371) have provided (necessary and) sufficient conditions for the existence of a non-trivial steady-state solution, we now have to derive for the present CD OLG model with nonlinearly regenerating resource stock restrictions on the nature of the interaction between preferences, technology, and natural growth which guarantee the existence of a non-trivial steady-state solution. At first sight such existence conditions could be simply borrowed from ILA models (as, e.g., in Clark, 1990; Neher, 1990; Tahvonen, 1991; Beltratti et al, 1998). However, Koskela et al (2002, 498) rightly point out that the finite life of overlapping generations, implying inter alia resource-stock trade between subsequent generations in intertemporal general equilibrium, 'brings a striking difference to the results of traditional analyses' and requires therefore an independent analysis even if ultimately under specific dynamic structures the existence conditions in both model structures turn out to be similar.

With the slopes of the RR-curve (9.26) and the XX-curve (9.27) at hand, we are able to state sufficient conditions for the existence of a strictly positive, unique steady state (X, R).

Proposition 9.1 (Existence and uniqueness of non-trivial steady state). *If* $lim_{R\to 0}\left[(\mathrm{d}X/\mathrm{d}R)\,|_{R_{t+1}=R_t} - (\mathrm{d}X/\mathrm{d}R)|_{X_{t+1}=X_t}\right] > 0$, *a unique, non-trivial solution* $0 < R < R_{max}$ *of* $\mathrm{g}(R) = f(R)$ *satisfying* (9.17) *exists. For the parametric example of the logistic regeneration function*

$$\lim_{R\to 0}\left[\left.\frac{\mathrm{d}X}{\mathrm{d}R}\right|_{R_{t+1}=R_t} - \left.\frac{\mathrm{d}X}{\mathrm{d}R}\right|_{X_{t+1}=X_t}\right] = \mathrm{g}'(0) - \frac{\alpha_1\left[1+\mathrm{g}'(0)\right]^2}{(\alpha_1+\alpha_2\sigma)\left[1+\mathrm{g}'(0)\right]-\alpha_3}. \quad (9.28)$$

Proof. Since $\mathrm{g}''(R) < 0$, $g'(R)$ can be inverted such that we can define a unique $\hat{R} \equiv \mathrm{g}'^{-1}(\alpha_3/(\alpha_1+\alpha_2\sigma)-1)$. The strict concavity of $g(R)$ implies that the average yield of the natural capital is greater than the marginal yield: $\phi(R)$ is therefore strictly positive for positive R. Because by assumption $1+g'(R) > 0$, the numerator in (9.25) is positive for all positive R. Thus, $f(R) > 0, 0 < R < \hat{R}$ if $[1+g'(R_t)](\alpha_1+\alpha_2\sigma) \geq \alpha_3$. However, this inequality is implied by the derivative condition above. Given that, $lim_{R\to -\hat{R}} f(R) = +\infty$. Because $\mathrm{g}(R)$ is bounded, certainly $\exists\varepsilon > 0$ such that $f(\hat{R}-\varepsilon) - \mathrm{g}(\hat{R}-\varepsilon) > 0$. On the other hand, we already know that $f(0) - \mathrm{g}(0) = 0$. The derivative assumption in Prop. 9.1 implies that $\exists\varepsilon > 0$ such that $f(R) < \mathrm{g}(R) \Leftrightarrow f(R) - \mathrm{g}(R) < 0, \forall R \in B_\varepsilon(0)$. Given the continuity of $f(R)$ and $\mathrm{g}(R)$ on $(0, \hat{R})$, the intermediate-value theorem implies that $\exists R \in (0, \hat{R})$ such that $f(R) - \mathrm{g}(R) = 0 \Leftrightarrow f(R) = \mathrm{g}(R)$.

Clearly, $X = f(R) = \mathrm{g}(R) \leq R + \mathrm{g}(R)$, since $R \geq 0$. Thus, the upper constraint in (9.17) is fulfilled by the steady state solution. Finally, $R < R_{max}$ otherwise when $R \geq$

R_{max}, then $f(R) = f(R_{max}) = g(R) = g(R_{max}) = 0$ which contradicts $f(R_{max}) = R_{max} + g(R_{max}) = R_{max} > 0$.

In order to prove the uniqueness of the steady-state solution, monotonicity of $f(R)$ and $g(R)$ is needed. However, $g(R)$ is in $(0, R_{max})$ either monotonically increasing $[g'(R) \geq 0]$ or monotonically decreasing $[g'(R) < 0]$. Comparing (9.25) and (9.27), $lim_{R \to -\hat{R}} f(R) = +\infty$ implies $lim_{R \to -\hat{R}} f'(R) = +\infty$. Since the derivative assumption in Prop. 9.1 implies $[1 + g'(R_t)](\alpha_1 + \alpha_2 \sigma) > \alpha_3$, the denominator of (9.27) is certainly strictly positive. The numerator of (9.27) remains positive for $R < \hat{R}$ because all terms are positive except $\sigma g''(R) R$, and this term decreases for $R \leq \hat{R} - \varepsilon$, $\varepsilon > 0$, if $g'''(R) \geq 0$ which is surely the case for the parametric example of the logistic regeneration function. Thus, $f(R)$ is monotonic for $R \in (0, \hat{R})$.

For the parametric example of the logistic regeneration function there is an algebraic solution for the steady state resource stock as follows:

$$R = \frac{\Omega \{\Pi [(\alpha_2 - \alpha_1)\sigma] + 2[\alpha_2 \Pi \sigma - \alpha_1 - \alpha_2 \sigma - \alpha_3] + \alpha_3\}}{4\Omega^2 [(\alpha_2 - \alpha_1)\sigma]} - \frac{\sqrt{\Omega^2 \{\Pi [(\alpha_2 - \alpha_1)\sigma] + 2[\alpha_2 \Pi \sigma - \alpha_1 - \alpha_2 \sigma - \alpha_3] + \alpha_3\}^2 - 8\Omega^2 B}}{4\Omega^2 [(\alpha_2 - \alpha_1)\sigma]},$$

with $B \equiv [(\alpha_2 - \alpha_1)\sigma][\Pi(\alpha_2 \Pi \sigma - \alpha_1 - \alpha_2 \sigma - \alpha_3) + \alpha_3]$. □

To facilitate the economic interpretation of the existence condition, a geometrical illustration using Fig. 9.1 or 9.2 is in order. The condition in Prop. 9.1 requires that the slope of the *RR*-curve at the origin (= growth potential of the resource stock) be larger than the slope of the *XX*-curve at the origin. Economically interpreted the existence condition says that the growth potential (in terms of sustained harvesting) of the resource stock is larger than the increase of economic general-equilibrium demand for sustained resource harvesting due to a marginally increased resource stock. As the positive slope of the *XX*-curve in Figs. 9.1 and 9.2 indicates, the demand for sustained resource harvesting in general equilibrium depends positively on the resource stock. This positive dependence is, on the one hand, due to the wealth effect of higher natural capital which means that households' consumption demand for the man-made product and—weighted by the production elasticity of resource harvesting—the production input demand for resource harvesting increases and, on the other hand, due to the portfolio-reallocation effect of higher natural capital to more man-made capital investment and—weighted by the production elasticity of resource harvesting—to more natural production input. In the contrary case, if a marginal change of the resource stock induced a larger change of the general-equilibrium harvesting demand than that of nature's harvesting supply, only zero production could be a steady-state solution.

9.5 Stability of the Steady States

As usual, the local-stability properties of the steady-state solutions are investigated by calculating the eigenvalues of the Jacobian matrix at the steady states, $J(X,R)$, defined by:

$$J(X,R) = \begin{pmatrix} \frac{\partial X_{t+1}}{\partial X_t} & \frac{\partial X_{t+1}}{\partial R_t} \\ \frac{\partial R_{t+1}}{\partial X_t} & \frac{\partial R_{t+1}}{\partial R_t} \end{pmatrix}. \tag{9.29}$$

The partial derivatives in (9.29) are calculated by partial differentiation of the difference equations (9.18) and (9.19) with respect to resource harvesting and the resource stock and evaluated at the steady states:

$$\frac{\partial X_{t+1}}{\partial X_t} = -g''(R)\left[\frac{g(R)}{1+g'(R)}\right] + \alpha_3^{-1}\left[1+g'(R)\right](\alpha_1+\alpha_2\sigma) > 0, \tag{9.30a}$$

$$\frac{\partial X_{t+1}}{\partial R_t} = g(R)g''(R) - \frac{\alpha_1}{\alpha_3}\left[1+g'(R)\right]\left\{1+g'(R)+\sigma g''(R)R\right\}, \tag{9.30b}$$

$$\frac{\partial R_{t+1}}{\partial X_t} = -1, \tag{9.30c}$$

$$\frac{\partial R_{t+1}}{\partial R_t} = 1+g'(R) > 0. \tag{9.30d}$$

Let us denote the trace of the Jacobian $J(X,R)$ by $\mathrm{tr}J(X,R)$, the determinant of this Jacobian by $\det J(X,R)$, the discriminant of this Jacobian by $\Delta J(X,R) \equiv \mathrm{tr}J(X,R)^2 - 4\det J(X,R)$ and let the eigenvalues of this Jacobian be ${}_1\psi_2 = \frac{1}{2}\left[\mathrm{tr}J(X,R) \pm \sqrt{\Delta J(X,R)}\right]$.

From Galor (1992, 1383) or Azariadis (1993, 63-67) we know: If the discriminant and the determinant of the Jacobian at the trivial steady state are strictly positive, if the trace is larger than 2 and if $1-\mathrm{tr}J(0,0)+\det J(0,0) > 0$, both eigenvalues are larger than one which means that the equilibrium dynamics is asymptotically unstable $(\psi_1 > 1, \psi_2 > 1)$.

With this knowledge at hand, we can determine the stability properties of the trivial and non-trivial steady state.

Proposition 9.2. *The equilibrium dynamics* (9.18) *and* (9.19) *evaluated at the trivial steady state is asymptotically unstable.*

Proof. Using (9.30), simple calculation yields:

$$\mathrm{tr}J(0,0) = 1+g'(0) + \left[\frac{\alpha_1+\alpha_2\sigma}{\alpha_3}\right]\left[1+g'(0)\right] > 2, \tag{9.31}$$

$$\det J(0,0) = \frac{\alpha_1}{\alpha_2}\sigma\left[1+g'(0)\right]^2 > 0, \tag{9.32}$$

$$\Delta J(0,0) = \left[1+g'(0)\right]^2\alpha_3^{-1}\left[(\alpha_3-\alpha_2\sigma)^2+\alpha_1^2+2\alpha_1(\alpha_3+\alpha_2\sigma)\right] > 0. \tag{9.33}$$

Thus,

$$1-\operatorname{tr}J(0,0)+\det J(0,0)>0 \tag{9.34}$$

$$\Longleftrightarrow \alpha_3+\alpha_2\sigma\left[1+g'(0)\right]^2>\left[1+g'(0)\right](\alpha_1+\alpha_2\sigma+\alpha_3) \tag{9.35}$$

which exactly equals the existence condition of Prop. 9.1. □

Proposition 9.3. *The non-trivial steady state resource stock R and the corresponding stationary resource yield X represent a non-oscillating saddle point.*

Proof. The eigenvalues of the Jacobian (9.29) ${}_1\psi_2=\frac{1}{2}\left[\operatorname{tr}J(X,R)\pm\sqrt{\Delta J(X,R)}\right]$ at the non-trivial steady state are real and distinct if the discriminant of the Jacobian defined by

$$\begin{aligned}\Delta J(X,R)=&\left[1+g'(R)\right]^{-2}\left\{-g''(R)g(R)+\left[1+g'(R)\right]^2\left[\frac{\alpha_1+\alpha_2\sigma-\alpha_3}{\alpha_3}\right]\right\}^2+\\&-4g''(R)g(R)+4\left[1+g'(R)\right]\left(\frac{\alpha_1}{\alpha_3}\right)\left[1+g'(R)+g''(R)R\right]\end{aligned} \tag{9.36}$$

is positive. Comparing this discriminant expression with the numerator of (9.27), the former is certainly positive if the latter is. From the proof of the uniqueness of the steady state solution, we know that (9.27) is positive.

The eigenvalues are strictly positive if $\operatorname{tr}J(X,R)>0$ and $\det J(X,R)>0$, since $\psi_1+\psi_2=\operatorname{tr}J(X,R)>0$ and $\psi_1\psi_2=\det J(X,R)>0$. Now

$$\operatorname{tr}J(X,R)=\frac{-g''(R)g(R)}{\left[1+g'(R)\right]}+\left[1+g'(R)\right]\left[\frac{\alpha_1+\alpha_2\sigma+\alpha_3}{\alpha_3}\right]>0, \tag{9.37}$$

$$\det J(X,R)=\left(\frac{\alpha_2\sigma}{\alpha_3}\right)\left[1+g'(R)\right]^2-\left(\frac{\alpha_1\sigma}{\alpha_3}\right)g''(R)\left[1+g'(R)\right]R>0. \tag{9.38}$$

From Galor (1992, lemma A1, 1383) $\psi_1>1, 0<\psi_2<1$, iff $\operatorname{tr}J(X,R)>0$, $\det J(X,R)>0$, $\Delta J(X,R)>0$, and $1-\operatorname{tr}J(X,R)+\det J(X,R)<0$:

$$\begin{aligned}&1-\operatorname{tr}J(X,R)+\det J(X,R)=\\&1+\left\{\frac{g''(R)g(R)}{\left[1+g'(R)\right]}\right\}-\left[1+g'(R)\right]\left[\frac{\alpha_1+\alpha_2\sigma+\alpha_3}{\alpha_3}\right]+\\&+\left(\frac{\alpha_2\sigma}{\alpha_3}\right)\left[1+g'(R)\right]^2-\left(\frac{\alpha_1\sigma}{\alpha_3}\right)g''(R)\left[1+g'(R)\right]R.\end{aligned} \tag{9.39}$$

We already know that the trace and the determinant of the Jacobian evaluated at the non-trivial steady state are strictly positive. In order to show that $1-\operatorname{tr}J(X,R)+\det J(X,R)<0$, we make use of the XX- and RR-curve and their derivatives derived in (9.26) and (9.27). It is easy to verify that $1-\operatorname{tr}J(X,R)+\det J(X,R)<0\Leftrightarrow(\mathrm{d}X/\mathrm{d}R)|_{X_{t+1}=X_t}>(\mathrm{d}X/\mathrm{d}R)|_{R_{t+1}=R_t}$. Therefore, the non-trivial steady state is a stable, non-oscillating saddle point if the XX-curve cuts the RR-curve from below.

Because $(\mathrm{d}X/\mathrm{d}R)|_{X_{t+1}=X_t} > 0$, this is certainly the case when $(\mathrm{d}X/\mathrm{d}R)|_{R_{t+1}=R_t} < 0$ (as in Fig. 9.2). Otherwise (as in Fig. 9.1), this stability condition is implied by $(\mathrm{d}X/\mathrm{d}R)|_{X_{t+1}=X_t} > 0$ and the existence assumption of Prop. 9.1. □

The analytical information provided by the Props. 9.1-9.3 is summarized by Figs. 9.1 and 9.2. Clearly, the non-trivial steady-state solution (X, R) is exactly there where the stationary-state curves intersect in the first orthant of the $X-R$ space. In Fig. 9.1, the XX-curve cuts the RR-curve in the area where the slope of the RR-curve is positive: the steady state resource stock is such that the natural own rate of return and hence the real interest rate are positive. In Fig. 9.2, the steady-state solution exhibits a negative natural own rate of return and a negative real interest rate. Both solutions are compatible with Prop. 9.1.

The instability of the trivial steady state and the saddle-point stability of the non-trivial solution are illustrated by Figs. 9.1 and 9.2. An inspection of the stationary-state curves in both diagrams reveals that the XX-curve cuts the RR-curve at the non-trivial steady state resource stock from below: as shown by the proof to Prop. 9.3, saddle-point stability at the non-trivial steady state demands that the derivative of the XX-curve is greater than the derivative of the RR-curve at the steady state resource stock (R in Figs. 1 and 2).[3]

9.6 The Saddle Paths Converging to the Non-Trivial Steady State

The proof of Prop. 9.3 reveals that the eigenvalue ψ_1 of the Jacobian evaluated at R can be rejected since otherwise the equilibrium dynamics would become asymptotically unstable. Only the second eigenvalue $0 < \psi_2 < 1$ allows for a non-explosive equilibrium dynamics. But the exclusion of ψ_1 implies that the equilibrium dynamics is only saddle-path ('knife-edge') stable.

A linear approximation of the equilibrium dynamics evaluated at the saddle point (X, R) reads as follows:

$$\begin{pmatrix} X_{t+1} \\ R_{t+1} \end{pmatrix} = (J(X,R)) \begin{pmatrix} X_t \\ R_t \end{pmatrix} + (I - J(X,R)) \begin{pmatrix} X \\ R \end{pmatrix}. \tag{9.40}$$

The general solution of this linearly approximated difference-equation system takes the following form:

$$X_t = X + \kappa e_X \psi_2^t, \tag{9.41}$$

$$R_t = R + \kappa e_R \psi_2^t, \tag{9.42}$$

with $\kappa \neq 0$ and $\binom{e_X}{e_R}$ being the solution of the following matrix equation:

[3] The phase-diagrams in Koskela et al (2002, 505) look very similar to ours in Figs. 9.1 and 9.2. This fact is not really surprising because our log-linear CD model with man-made capital is at least dimensionally equivalent to Koskela et al's model without man-made capital but with quasi-linear utility function.

$$(J(X,R))\begin{pmatrix} e_X \\ e_R \end{pmatrix} = \psi_2 \begin{pmatrix} e_X \\ e_R \end{pmatrix}. \tag{9.43}$$

It is easy to see that one admissible solution of (9.43) reads as follows:

$$\begin{pmatrix} e_X \\ e_R \end{pmatrix} = \begin{pmatrix} J_{22}(X,R) - \psi_2 \\ 1 \end{pmatrix} \text{ with } J_{22}(X,R) = 1 + g'(R). \tag{9.44}$$

Now, consider (9.42) at $t = 0$ and put $e_R = 1$ from (9.44). This yields:

$$\kappa = R_0 - R. \tag{9.45}$$

Inserting (9.44) and (9.45) in (9.41) and (9.42), one is prepared to state the following

Proposition 9.4. *Let $J_{22}(X,R)$ be given from (9.44) and let ψ_2 be given from the proof to Prop. 9.3. A linear approximation of the equilibrium dynamics of the natural-resource variables evaluated at the saddle point (X,R) takes the following form:*

$$R_{t+1} = R(1-\psi_2) + \psi_2 R_t, \text{ given } R_0 > 0, \tag{9.46a}$$

$$X_t = X + [J_{22}(X,R) - \psi_2](R_t - R), \tag{9.46b}$$

$$K_{t+1} = \left[\frac{\alpha_3 X_{t+1}}{1+g'(R_{t+1})}\right] X_t^{\alpha_1 - 1} K_t^{\alpha_3}, \text{ given } K_0 > 0. \tag{9.46c}$$

With this equilibrium dynamics near the non-trivial steady state at hand, we will now investigate whether there are parameter combinations of the dynamical system at all which allow both for accumulation of man-made capital and time-stationary stocks of the renewable resource, i.e. for a sustainable growth path as defined by Mourmouras (1991, 589) approaching: $K = \{(\alpha_3 X^{\alpha_1})/[1+g'(R)]\}^{1/(1-\alpha_3)} > K_0$, whereby $R_t = R_0, \forall t$ with $R_0 > 0$.

Corollary 9.1 (Feasibility of sustainable growth). *Man-made capital accumulation and the equality of resource stocks across generations are incompatible with decentralized individual optimization and intertemporal market clearing except when $R_0 = R$.*

Proof. In view of the saddle-path dynamics depicted in Prop. 9.4 man-made capital accumulation $(K_0 < K)$ and intergenerationally equal natural capital $(R_t = R_0, \forall t)$ are compatible only if by accident $R_0 = R$. When however $R_0 \neq R$, intergenerational resource equality would demand $\psi_2 = 1$ which is inconsistent with Prop. 9.3. □

While sustainable growth is not feasible in an OLG model with logistically regenerating resources, in the steady state the renewable resource is time-stationary and ecologically sustainable by definition. The same is also true with respect to the utility of younger generations in the steady state. In the long run both notions of sustainability are fulfilled in the CD OLG model of this chapter.

9.7 Conclusions

This chapter focused both on the existence and dynamic stability of steady states in an intertemporal OLG market equilibrium model with a logistically regenerating resource stock. Steady states (stationary states) are by definition economically and ecologically sustainable. However, steady state sustainability is by no means self-evident. Finite lives of agents, generations' overlap, intergenerational selfishness, and trade of resource stocks among generations in perfectly competitive markets bring about in spite of log-linear intertemporal preferences and a CD production function a strong tendency of the OLG economy to contract towards the trivial steady state. To counter this contraction tendency the natural regeneration rate near a zero resource stock has to be sufficiently high in comparison to the stationary-state general equilibrium propensity to use the resource as production input (evaluated at the trivial steady state).

Given this existence condition the asymptotic instability of the trivial steady state and the saddle-path stability of the non-trivial steady state follow logically. Hereby the log-linearity of the preferences and the Cobb-Douglas form of the production function play an important role.

With the existence and dynamic stability of the non-trivial steady state we are also ensured that in the long run both economic and ecologic sustainability is feasible in an unhampered market economic with logistic (non-linear) regeneration. On the other hand, ecologically sustainable growth is not compatible with unhampered market competition when the resource stock over time has to remain strictly at its initial value. If the initial resource stock is less than the steady state value and a rise of the resource stock over time is allowed a sustainable growth path from an initially low man-made capital stock towards a higher steady state value exists even under logistic regeneration. If however the initial resource stock is higher and the man-made capital stock is lower than the respective steady state values (the realistic case), ecological sustainability neither in the strict nor in the weaker form is attainable in the logistic model.

In this chapter we neglect for the sake of simplicity harvest costs—an assumption which precludes the analysis of the impacts of harvest cost shocks in our CD OLG model. To enable the impact analysis we introduce stock dependent harvest costs into our model economy in the following chapter.

References

Azariadis C (1993) Intertemporal Macroeconomics. Blackwell, Oxford

Baumol WJ, Oates WE (1988) The theory of environmental policy, 2nd edn. Cambridge University Press, Cambridge, New York

Beltratti A, Chichilnisky G, Heal G (1998) Sustainable Use of Renewable Resources, Kluwer, Dordrecht, chap Interpreting Sustainability

Clark CW (1990) Mathematical Bioeconomics: The Optimal Management of Renewable Resources, 2nd edn. Wiley, NY

Diamond P (1965) National debt in a neoclassical growth model. American Economic Review 55:1126–1150
Farmer K (1989) Balance sheets and budget contraints in discrete-time macro-models: a note. European Journal of Political Economy 5:533–542
Farmer K (2000) Intergenerational natural–capital equality in an overlapping–generations model with logistic regeneration. Journal of Economics 72(2):129–152
Foley DK (1975) On two specifications of asset equilibrium in macroeconomic models. Journal of Political Economy 83:303–324
Galor O (1992) A two–sector overlapping–generations model: A global characterization of the dynamical system. Econometrica 60(6):1351–1386
Galor O, Ryder HE (1989) Existence, uniqueness, and stability of equilibrium in an overlapping–generations model with productive capital. Journal of Economic Theory 49:360–375
Hotelling H (1931) The economics of exhaustible resources. Journal of Political Economy 39:137–175
Koskela E, Ollikainen M, Puhakka M (2002) Renewable resources in an overlapping generations economy without capital. Journal of Environmental Economics and Management 43(3):497–517
Mourmouras A (1991) Competitive equilibria and sustainable growth in a life–cycle model with natural resources. Scandinavian Journal of Economics 93(4):585–591
Neher P (1990) Natural Resource Economics: Conservation and exploitation. Cambridge University Press, Cambridge, New York
Olson LJ, Knapp KC (1997) Exhaustible resource allocation in an overlapping generations economy. Journal of Environmental Economics and Management 32:277–292
Tahvonen O (1991) On the dynamics of renewable resource harvesting and pollution control. Environmental and Resource Economics 1:97–117
WCED (1987) Our Common Future. Oxford University Press, Oxford

Part V
Shocks to Harvest Technology and Natural Regeneration

Chapter 10
Resource Use with Physical Harvest Costs

10.1 Introduction

In Chaps. 8 and 9 we implicitly assumed that resource harvest does not, unlike commodity production, require inputs such as labor or capital. This is, however, an unrealistic simplification and this is why resource harvest costs are commonly found in sectoral models (for an overview, see Clark, 1990; Neher, 1990; Brown, 2000). In these models, harvest costs typically depend not only on the harvest volume as in Chaps. 6 and 7 but also on the resource stock. Applied to fisheries, harvest costs depend on the size of catch (Heaps and Neher, 1979) and on the available fish stock, following the general wisdom 'the more fish the easier to catch' (Smith, 1968; Tahvonen and Kuuluvainen, 2000). The model of this chapter incorporates therefore harvest costs which depend inversely on the resource stock. Since 'most fisheries in the world do not pay a wage rate but pay a share of the catch' (Brown, 2000, 881), we assume moreover that harvest costs are accountable in resource units and thus affect the size of the stock. Both the assumptions of stock dependent harvest costs and of physical accountability are a clear difference to the harvest cost function employed in Chaps. 6 and 7.

Natural resources form an essential input in the production of not one, but many goods. Thus, it is useful to investigate the resource technology shocks not in a partial but in a general equilibrium framework. However, the overwhelming majority of general equilibrium models with renewable natural resources have avoided the explicit modeling of harvest technologies and associated costs, an approach we also followed in Chaps. 8 and 9. Thus, the notion of harvest costs still is rare in intertemporal general equilibrium models of the ILA type (Krutilla and Reuveny, 2004), and completely missing within the OLG type. As seen from partial equilibrium analysis, this simplification is not only manifestly unrealistic but also precludes the investigation of the effects of harvest cost shocks. The omission of harvest cost in general equilibrium models is based on the implicit assumption of costs as a simple mark–up on the harvest price (Berck, 1981).

K. Farmer and B. Bednar-Friedl, *Intertemporal Resource Economics: An Introduction to the Overlapping Generations Approach*, DOI 10.1007/978-3-642-13229-2_10,

The present chapter offers therefore a conceptual integration of stock dependent harvest costs into our OLG model. As a result of this fundamental change in resource technology, an analysis of the properties of steady state solutions is necessary. This analysis is necessary since on account of harvest costs, a non-trivial steady state might not exist if private savings are insufficient to sustain capital and resource accumulation in the long run. Galor and Ryder (1989) show in an OLG model with man-made capital only that strengthened Inada conditions are necessary to ensure the existence of a non-trivial steady state. Obviously, in our OLG economy this problem is reinforced since stock-dependent harvest costs are particularly high if the resource stock is small. Moreover, the introduction of harvest costs can cause multiple equilibria (see Krutilla and Reuveny, 2004, in an ILA context), raising the question of uniqueness in our model. Regarding intergenerational efficiency, it remains to be seen whether the introduction of non-negligible harvest costs changes the efficiency insights gained so far in the previous chapters. Thus, the questions of existence, uniqueness and stability of a non-trivial steady state equilibrium with stock-dependent harvest costs as well as its efficiency will be carefully addressed in this chapter. In our existence and stability analysis, we closely follow the approach developed for two–sector OLG models by Galor (1992).

This chapter is structured as follows. The following section contains a description of the model and the derivation of the intertemporal equilibrium dynamics. After deriving the intertemporal equilibrium loci, we derive the conditions for a non-trivial steady state to exist, followed by the characteristics of a long-run intergenerationally efficient solution. The next step is to analyze the stability and uniqueness of steady state solutions.

10.2 An OLG model with Harvest Costs

The present model is an extension of the model presented in Chap. 9 in which harvest was assumed to be costless. For incorporating harvest costs, two options are available. The first is based on partial equilibrium models where harvesting requires labor (or effort) as input (see, e.g. Krutilla and Reuveny, 2004; Elíasson and Turnovsky, 2004). Usually, this approach is incorporated together with the well–known Schaefer (1954) harvest function, a functional specification popular in mostly partial equilibrium fishery models (see, e.g. Brown, 2000; Clark, 1990). Extending this harvest cost formulation to a general equilibrium framework implies that labor needs to be split optimally between two competing purposes, namely resource harvest and commodity production. This considerably complicates the model since labor becomes now an endogenous variable.

An alternative approach, with similar qualitative implications but considerably simpler analysis, is to assume that resource harvest causes costs to the resource stock (physical harvest costs). Under such an assumption, the resource stock is reduced by some gross harvest level of which only a fraction can be used as resource input in commodity production. Comparable specifications were also used by Armstrong and Sumaila (2001) and Escapa and Prellezo (2003) who both assume that the resource harvest technology alters the regeneration rate of the resource stock. Our specification of harvest costs is analogous to the introduction of adjustment costs into capital accumulation by Hayashi (1982) where gross investment turn only partly into capital. Hence, we will deploy this latter approach in this and the following chapter. For a brief outline of the model with endogenous labor supply, we refer to Appendix A.

10.2.1 Dynamics of the Resource Stock

The economy is endowed with a renewable resource stock, R_t, which for expositional purposes we shall identify as being a forest or a freshwater fishery. Property rights for the natural resource are fully specified and enforced without cost. In particular, the resource stock is owned by the households and can be either harvested or saved, as an alternative investment to man-made capital, to transfer income into the retirement period.

As in the previous chapter, we assume that resource regeneration is governed by the logistic function, but in slightly different notation:[1]

$$g(R_t) \equiv \pi \left(R_t - \frac{R_t{}^2}{R_{\max}} \right), \tag{10.1}$$

where π denotes the regeneration rate, or increase in units of biomass, and $R_{\max}$ the carrying capacity. In the absence of harvesting, the resource stock tends towards $R_{\max}$.

Costly resource harvest is represented through a resource harvest process $h(R_t)X_t$ where X_t is the harvest volume and the harvest cost function $h(R_t)$ which depends inversely on the size of the resource stock: $h' < 0$, $h'' > 0$ and $h(0)$ positive but finite. Accordingly, the scarcer the resource the more costly it is to harvest. Harvest costs in our model can be best understood as costs in units of the resource stock that accrue additionally to the harvest volume, either as unintended side–effects of resource harvest such as damage to livestock and forest stands, or as costs of employment effort paid in resources, not money. The harvest cost function is specified by

$$h(R_t) = 1 + \frac{\lambda}{\rho + R_t}, \tag{10.2}$$

with the harvest cost parameters $\lambda > 0$ and ρ small (but positive). Without harvest costs, we had $\lambda = 0$. Thus, λ reflects the effort or difficulty necessary to harvest one unit. For a fishery, the value of λ can be understood as the catchability of fish while for a forest it could represent the accessibility or location of the forest stand.

[1] The current version can be easily transformed into the earlier version by setting $\Omega \equiv \pi / R_{max}$.

Incorporating resource harvest and natural resource growth gives for the net regeneration function:

$$R_{t+1} = R_t + \mathrm{g}(R_t) - \mathrm{h}(R_t)X_t. \tag{10.3}$$

Accordingly, the resource stock in (10.3) is reduced by gross harvest $\mathrm{h}(R_t)X_t$ but only X_t can be actually sold on the market. The case of costless harvest as in Chap. 9 corresponds to $\mathrm{h}(R_t) = 1$.

10.2.2 Household and Firm Optimization

As in the chapters before, the representative consumer's intertemporal utility depends on consumption during the working period, c_t^1, and consumption during the retirement period, c_{t+1}^2, the latter being discounted by the time discount factor $0 < \beta < 1$. The commodity can both be consumed and invested, and the commodity in period t serves as the numeraire. For simplicity, the representative household's preferences are represented by a log–linear intertemporal utility function:

$$u = u(c_t^1, c_{t+1}^2) = ln\ c_t^1 + \beta\ ln\ c_{t+1}^2. \tag{10.4}$$

In maximizing intertemporal utility (10.4), the young household is constrained by a budget constraint in each period of life. Assuming that the resource stock is acquired at the beginning of the period and that working time is normalized to one, income when young is gained from employment and selling of the resource harvest X_t, and is spent on consumption c_t^1. Furthermore, for transferring income to their retirement period, young households save in terms of man-made capital K_{t+1} and in terms of the natural resource R_t:

$$p_t\ R_t + c_t^1 + K_{t+1} = w_t + q_t X_t, \tag{10.5}$$

where w_t denotes real wage, and q_t the price of resource harvest. The resource stock is bought at the beginning of the period and the resource harvest is sold simultaneously at a one–period forward market to the representative firm where it is used as an input to commodity production. This specification of resource stock holding follows a beginning–of–period notion (as in in Chap. 9) in contrast to the end–of–period notion employed in Koskela et al (2002).

From saving, the household gains income in the retirement period, where r_{t+1} denotes the interest factor of man-made capital and p_{t+1} the price of the resource stock. Thus, revenues when old derive from sale of the capital stock and of the resource stock to the younger generation at price p_{t+1} and are spent on consumption c_{t+1}^2 only:

$$c_{t+1}^2 = r_{t+1}K_{t+1} + p_{t+1}R_{t+1}. \tag{10.6}$$

As argued above, the resource stock is private property of the households and therefore the dynamics of the resource stock (10.3) form the third constraint to household optimization.

At an interior solution of the household's optimization problem, the following FOCs hold. They require that the intertemporal marginal rate of substitution equals the interest factor (10.7), that the net return on resource harvest equals the discounted price of the resource stock (10.8), and that the returns on man-made capital (r_{t+1}) and the renewable resource stock ($p_{t+1}/p_t\,(1+\mathrm{g}'(R_t)-\mathrm{h}'(R_t)X_t)$) are balanced (10.9):

$$\frac{c_{t+1}^2}{\beta c_t^1} = r_{t+1}, \tag{10.7}$$

$$\frac{q_t}{\mathrm{h}(R_t)} = \frac{p_{t+1}}{r_{t+1}}, \tag{10.8}$$

$$r_{t+1} = \frac{p_{t+1}}{p_t}\Big(1+\mathrm{g}'(R_t)-\mathrm{h}'(R_t)X_t\Big). \tag{10.9}$$

Deriving the intertemporal budget constraint from (10.5) and (10.6), taking account of (10.7)–(10.9), yields the consumption of the young in period t:

$$c_t^1 = \gamma\left\{w_t + \frac{q_t}{\mathrm{h}(R_t)}\Big(\mathrm{g}(R_t)-\mathrm{g}'(R_t)R_t+\mathrm{h}'(R_t)X_tR_t\Big)\right\}, \tag{10.10}$$

with $\gamma \equiv 1/(1+\beta)$. The consumption of the old is determined by (10.6) for t taking account of (10.8) and (10.9):

$$c_t^2 = r_tK_t + \frac{q_t}{\mathrm{h}(R_t)}\left(1+\mathrm{g}'(R_t)-\mathrm{h}'(R_t)X_t\right)R_t. \tag{10.11}$$

The firm is assumed to behave competitively and to maximize profits given output and input prices. Output Y_t is produced according to a constant returns to scale CD production function with labor N_t^y, capital K_t, and resource harvest X_t as inputs, where $Y_t = (X_t)^{\alpha_1}(N_t^y)^{\alpha_2}(K_t)^{\alpha_3}$, $\alpha_1+\alpha_2+\alpha_3 = 1$ and $0<\alpha_i<1, i = 1,2,3$ denote the constant production elasticities of the resource harvest, labor, and capital services, respectively. Per–period profit maximization yields the FOCs in the usual manner:

$$q_tX_t = \alpha_1Y_t, \qquad w_tN_t^y = \alpha_2Y_t, \qquad r_tK_t = \alpha_3Y_t. \tag{10.12}$$

All markets are assumed to clear every period, i.e. the markets for the resource stock (traded between households of different age), for resource harvest (traded between young households and the firm), for labor (where supply L_t is normalized to one, i.e. $N_t^y = 1$), and for man-made capital. Commodity market clearing coincides again with Walras' Law and is therefore redundant:

$$(X_t)^{\alpha_1}(K_t)^{\alpha_3} = c_t^1 + c_t^2 + K_{t+1}. \tag{10.13}$$

10.2.3 Intertemporal Equilibrium Dynamics

As in Chaps. 8 and 9, the intertemporal equilibrium dynamics can be reduced to a two–dimensional system in R_t and X_t by using the goods market clearing condition and the household's and firm's first order conditions. Similar to the model without harvest costs, a relationship for K_{t+1} can be derived from (10.13) by taking account of (10.10), (10.11), and (10.12). Another relationship for K_{t+1} can be derived from (10.8), (10.9), and (10.12). Equating for K_{t+1}, and taking account of (10.12), yields the equation of motion for resource harvest:

$$X_{t+1} \equiv \Phi^X(R_t, X_t) = \frac{(1+\mathrm{g}'(R_{t+1}))\,\mathrm{A}(R_t,X_t)}{\alpha_3\,\mathrm{h}(R_{t+1}) + \mathrm{h}'(R_{t+1})\mathrm{A}(R_t,X_t)}, \tag{10.14}$$

with $\mathrm{A}(R_t,X_t) \equiv [(\alpha_1+\alpha_2\sigma)\mathrm{h}(R_t) + \alpha_1\sigma\mathrm{h}'(R_t)R_t]\,X_t - \alpha_1\,[\gamma\Phi(R_t)+1+\mathrm{g}'(R_t)]\,R_t$, where $\Phi(R_t) \equiv \mathrm{g}(R_t)/R_t - \mathrm{g}'(R_t)$ and $\sigma \equiv 1-\gamma$.

In order to guarantee that $X_{t+1} \geq 0$ and $R_{t+1} \geq 0, \forall t$, resource harvest has to be bounded from above and below: $X_t \in [X_{\min}(R_t), X_{\max}(R_t)]$, $\forall R_t \in [0, R_{\max}], \forall t$. Let $X_t \geq X_{\min}(R_t), \forall t$. To ensure that in (10.14) $X_{t+1} \geq 0, \forall t$,

$$X_{\min}(R_t) \equiv \frac{\alpha_1 R_t\,[\gamma\Phi(R_t)+1+\mathrm{g}'(R_t)]}{(\alpha_1+\alpha_2\sigma)\mathrm{h}(R_t)+\alpha_1\sigma\mathrm{h}'(R_t)R_t},$$

with $(\alpha_1+\alpha_2\sigma)\mathrm{h}(R_t)+\alpha_1\sigma\,(\mathrm{h}'(R_t))\,R_t > 0$. According to the definition of the lower bound $X_t = X_{\min}(R_t)$, we have $\mathrm{A}(R_t,X_t)=0$ which corresponds also to the case that man-made capital is not required as an input to production. Since $(\alpha_1+\alpha_2\sigma)\mathrm{h}(R_t)+\alpha_1\sigma\,(\mathrm{h}'(R_t))\,R_t > 0$, it follows that $X_{\min}(R_t) > 0, \forall R_t \in (0, R_{\max}]$. For $R_{t+1} \geq 0, \forall t$, an upper bound for X_t follows from (10.15): $X_{\max}(R_t) \equiv R_t + \mathrm{g}(R_t)$.

The second equation of motion, $R_{t+1} \equiv \Phi^R(R_t,X_t)$, is identical to the stated natural growth function:

$$R_{t+1} \equiv \Phi(R_t,X_t) = R_t + \mathrm{g}(R_t) - \mathrm{h}(R_t)X_t. \tag{10.15}$$

While the derivation of (10.14) involves the capital stock implicitly, the explicit dynamics of the capital stock follow from:

$$K_{t+1} \equiv \Phi^K(R_t,X_t,K_t) = \frac{\alpha_3\,X_{t+1}\mathrm{h}(R_{t+1})X_t^{\alpha_1-1}K_t^{\alpha_3}}{(1+\mathrm{g}'(R_{t+1})-\mathrm{h}'(R_{t+1})X_{t+1})\,\mathrm{h}(R_t)}, \quad \text{for given } K_0 > 0. \tag{10.16}$$

A non-trivial perfect–foresight equilibrium is then a sequence $(R_t,X_t)_{t=0}^{\infty}$ such that (10.14)–(10.15) hold, where $R_t \in (0,R_{\max})$, $X_t \in (X_{\min}(R_t), X_{\max}(R_t))$, and $R_0 > 0$ given.

10.3 Derivation of Intertemporal Equilibrium Loci

To investigate the conditions for the existence and stability of a non-trivial steady state, we analyze the properties of the intertemporal equilibrium under perfect foresight by utilizing the notion of intertemporal equilibrium loci. This allows us, in contrast to the local analysis of the previous chapter, to conduct a global analysis of existence, stability, and uniqueness in the subsequent sections. The *RR*-locus, also referred to in the literature as isocline (see, e.g. Shone, 1997), is the geometrical place of all pairs (R_t, X_t) such that the resource stock is stationary ($R_t = \Phi^R(R_t, X_t)$ in (10.15)) while at the *XX*-locus the resource harvest is stationary ($X_t = \Phi^X(R_t, X_t)$ in (10.14)).[2]

Lemma 10.1 and 10.2 assure that the (functional) solutions to both loci are economically feasible and single valued such that the *RR*-locus is given by $X_t = \varphi(R_t)$ and the *XX*-locus by $X_t = \mu(R_t)$. For the *XX*-locus to exist and to be single valued, condition (10.17) is sufficient.[3]

Lemma 10.1 (*XX*-locus). *Let the regeneration function* $\mathrm{g}(R)$ *and the harvest cost function* $\mathrm{h}(R)$ *be specified by* (10.1) *and* (10.2)*, and let condition*

$$\alpha_3 \mathrm{h}(0) < \left(1 + \mathrm{g}'(0)\right)\left((\alpha_1 + \alpha_2\sigma)\mathrm{h}(0) - \alpha_1\right), \tag{10.17}$$

$\forall R_t \in (0, R_{\max}]$ *and for* ρ *near zero (but positive) hold. Then, there exists a unique* $X_t \in (X_{\min}(R_t), X_{\max}(R_t))$, $\forall R_t \in [0, R_{\max}]$, *such that* $(R_t, X_t) \in XX$, $X_t = \mu(R_t)$, *where* $\mu(R_t)$ *is a continuously differentiable function.*

Proof. See Appendix B.

Lemma 10.2 (*RR*-locus). *There exists a unique* $X_t \in [0, X_{\max}(R_t))$ *for all* $R_t \in [0, R_{\max}]$ *such that for* $(R_t, X_t) \in RR$, $X_t = \varphi(R_t) \equiv \mathrm{g}(R_t)/\mathrm{h}(R_t)$, *where* $\varphi(R_t)$ *is a continuously differentiable function.*

[2] Note that this loci, or isoclines, are different from the *RR*- and *XX*-curves of the previous chapter which were, strictly speaking, only defined in the steady state (X, R).

[3] Inspecting $X_t = \Phi^X(R_t, X_t)$ in (10.14) reveals that $\Phi^X(R_t, X_t)$ is a polynomial of order four in X_t which is analytically inaccessible. This is in contrast to the no harvest cost case where $\mathrm{h}(R_t) = 1$ and hence (10.14) is a second order polynomial in X_t. In the case without man-made capital, i.e. $\alpha_3 = 0$, we would have a non–polynomial solution for X_t. To identify the economically feasible and unique root, we apply the intermediate value theorem which yields condition (10.17).

10.4 Existence of Non-Trivial Steady State

By setting $R_{t+1} = R_t = R$ and $X_{t+1} = X_t = X$, for all t, a non-trivial steady state to the dynamic system (10.14) and (10.15) can be defined:

$$X = \varphi(R) \equiv \frac{\mathrm{g}(R)}{\mathrm{h}(R)}, \tag{10.18}$$

while the relationship $X = \mu(R)$ is defined implicitly by

$$\alpha_3 h(R)X = \left(1 + g'(R) - h'(R)X\right) \mathrm{A}(R,X). \tag{10.19}$$

For economically feasible values (i.e., $X \in (X_{\min}(R), X_{\max}(R))$ and $R \in [0, R_{\max}]$), system (10.18–10.19) defines the steady state(s).

There exists one trivial steady state $(R,X) = (0,0)$, but it will not be further investigated here, since the condition for its existence does not involve parameters of the harvest cost function and is thus similar to a corresponding OLG model without harvest costs (see Chap. 9) where a trivial steady state exists which is locally unstable.

Applying the methods developed by Galor (1992) for two–sector OLG models to prove the existence of a non-trivial steady state to the present model requires that (i) at the origin ($R = 0$), the RR-locus coincides with the XX-locus, (ii) at the carrying capacity $R_t = R_{\max}$, the RR-locus is below the XX-locus, and (iii) in the vicinity of the origin, the slope of the RR-locus is steeper than that of the XX-locus. Prop. 10.1 states precisely the condition for the existence of a non-trivial steady state.

Proposition 10.1 (Existence of non-trivial stationary state). *Let Condition* (10.17) *in Lemma 10.1 hold. Then, a non-trivial steady state (i.e. stationary state) solution exists if*

$$\lim_{R_t \to 0} \left.\frac{\mathrm{d}X_t}{\mathrm{d}R_t}\right|_{R_{t+1}=R_t} > \lim_{R_t \to 0} \left.\frac{\mathrm{d}X_t}{\mathrm{d}R_t}\right|_{X_{t+1}=X_t}.$$

But this is the case if $\mathrm{g}'(0) > \alpha_1 \left(1 + \mathrm{g}'(0)\right)^2 \left[(1 + \mathrm{g}'(0))(\alpha_1 + \alpha_2\sigma) - \alpha_3\right]^{-1}$.

Proof. We know that $\varphi(0) = \mu(0) = 0$, and $\varphi(R_{\max}) < \mu(R_{\max})$ since $\varphi(R_{\max}) = 0$ and $\mu(R_{\max}) > X_{\min}(R_{\max}) > 0$. In order to show that $\lim_{R_t \to 0} \mathrm{d}X_t/\mathrm{d}R_t|_{R_{t+1}=R_t} > \lim_{R_t \to 0} \mathrm{d}X_t/\mathrm{d}R_t|_{X_{t+1}=X_t}$ is sufficient for the existence of a non-trivial steady state, we have to investigate the slopes of the two loci for limes for $R_t \to 0$.

The slope of the RR-locus can be obtained by taking differentials of (10.15), with respect to X_t and R_t, considering $R_{t+1} = R_t$ and solving for $\mathrm{d}X_t/\mathrm{d}R_t$:

$$\left.\frac{\mathrm{d}X_t}{\mathrm{d}R_t}\right|_{R_{t+1}=R_t} = \frac{\mathrm{g}'(R_t) - \mathrm{h}'(R_t)X_t}{\mathrm{h}(R_t)}, \quad X_t = \varphi(R_t), \tag{10.20}$$

Proceeding similarly by differentiating (10.14'), considering $X_{t+1} = X_t$ yields for the slope of the XX-locus:

$$\left.\frac{\mathrm{d}X_t}{\mathrm{d}R_t}\right|_{X_{t+1}=X_t} = \frac{\partial \mathrm{RHS}/\partial R_t\,(R_t,X_t) - \partial \mathrm{LHS}/\partial R_t\,(R_t,X_t)}{\partial \mathrm{LHS}/\partial X_t\,(R_t,X_t) - \partial \mathrm{RHS}/\partial X_t\,(R_t,X_t)}, \quad X_t = \mu(R_t), \quad (10.21)$$

where $\partial \mathrm{LHS}/\partial X_t$ and $\partial \mathrm{RHS}/\partial X_t$ are known from (10.38)–(10.39) and

$$\frac{\partial \mathrm{LHS}(R_t,X_t)}{\partial R_t} = \alpha_3 \mathrm{h}'(R_{t+1})X_t\left(1+\mathrm{g}'(R_t) - \mathrm{h}'(R_t)X_t\right), \quad (10.22)$$

$$\frac{\partial \mathrm{RHS}(R_t,X_t)}{\partial R_t} = \left(1+\mathrm{g}'(R_{t+1}) - \mathrm{h}'(R_{t+1})X_t\right)\frac{\partial \mathrm{A}(R_t,X_t)}{\partial R_t} + \mathrm{A}(R_t,X_t)\times$$
$$\times\left(1+\mathrm{g}'(R_t) - \mathrm{h}'(R_t)X_t\right)\left[\mathrm{g}''(R_{t+1}) - \mathrm{h}''(R_{t+1})X_t\right]. \quad (10.23)$$

with

$$\partial \mathrm{A}(R_t,X_t)/\partial R_t = (\alpha_1+\alpha_2\sigma)\,\mathrm{h}'(R_t)X_t - \alpha_1\left[1+\mathrm{g}'(R_t) + \right.$$
$$\left. +\sigma\left(\mathrm{g}''(R_t) - \mathrm{h}''(R_t)X_t\right)R_t - \sigma \mathrm{h}'(R_t)X_t\right].$$

Evaluating (10.20) and (10.21) for $R_t \to 0$ yields:

$$\lim_{R_t\to 0}\varphi'(R_t) = \frac{\mathrm{g}'(0)}{\mathrm{h}(0)} > 0$$

$$\lim_{R_t\to 0}\mu'(R_t) = \frac{\alpha_1\left(1+\mathrm{g}'(0)\right)^2}{\left(1+\mathrm{g}'(0)\right)(\alpha_1+\alpha_2\sigma)\mathrm{h}(0) - \alpha_3\mathrm{h}(0)}$$

Because of Lemma 10.1, the denominator in $\lim_{R_t\to 0}\mu'(R_t)$ is positive. Thus, $\lim_{R_t\to 0}\varphi'(R_t) > \lim_{R_t\to 0}\mu'(R_t)$ is equivalent to the requirement that $\mathrm{g}'(0) > \alpha_1\left(1+\mathrm{g}'(0)\right)^2\left[\left(1+\mathrm{g}'(0)\right)(\alpha_1+\alpha_2\sigma) - \alpha_3\right]^{-1}$. Then, there exists an $\varepsilon > 0$, such that $\varphi(R_t) > \mu(R_t), \forall R_t \in (0,\varepsilon)$. Since furthermore $\varphi(\cdot)$ and $\mu(\cdot)$ are continuous on the interval $(0, R_{\max})$, it follows from the Intermediate Value Theorem that there exists an $R_t = R \in (0, R_{\max})$ such that $\varphi(R_t) = \mu(R_t)$. □

The slope condition in Prop. 10.1 requires that at the origin, the growth potential of the natural resource stock (represented by the slope of the *RR*-locus) has to be higher than the resource harvest demanded by the production sector (represented by the slope of the XX-locus). The demand for resource harvest depends positively on the available resource stock (since the slope of the XX-locus near the origin is positive) because with higher holdings of the resource stock the households' demand for the commodity increases. A higher resource stock leads to lower harvest cost since harvest cost depend inversely on the stock, partially offsetting the previous effect. Moreover, resource harvest increases because of a higher resource stock causing an asset relocation towards more man-made capital, the production of which demands higher resource harvesting. For a non-trivial steady state to exist, it is thus necessary that the resource harvest supply responds more to an increase in the resource stock than the harvest demand.

10.5 Long-Run Intergenerational Efficiency of Non-Trivial Steady State

Knowing that a non-trivial steady state exists, we investigate whether (or under which conditions) the steady state is long-run intergenerationally efficient. This is particularly relevant given the fact that the non-trivial steady state solutions may or may not be intergenerationally efficient in an OLG model with a renewable resource but without man-made capital (Koskela et al, 2002) and that the same holds in an OLG model with man-made capital and land (Rhee, 1991).

To investigate the efficiency of the non-trivial steady state, we first derive the FOCs for long-run intergenerational efficiency. A social planner maximizes utility of each individual living in the steady state ($u(c^1, c^2)$), subject to a certain level of welfare of the oldest generation ($u(c_0^2)$):

$$\max \ln c^1 + \beta \ln c^2$$

subject to:

$$\begin{aligned}
&(i) && \ln c_0^2 \geq \ln(c_0^2)^{\bullet}, \\
&(ii) && c^1 + c^2 + K = X^{\alpha_1} K^{\alpha_3}, \\
&(iii) && c^1 + c_0^2 + K = X^{\alpha_1} K_0^{\alpha_3}, \\
&(iv) && \mathrm{g}(R) = \mathrm{h}(R)X, \\
&(v) && R_0 + \mathrm{g}(R) = \mathrm{h}(R_0)X + R, \\
&(vi) && c_0^2, c^1, c_{t+1}^2 \geq 0,\ R, R_0 \geq 0,\ X \geq 0,\ K_o, K \geq 0,
\end{aligned}$$

where R_0 and K_0 are the resource and the capital stock owned by the initially old generation.

Setting up the Lagrangian

$$\begin{aligned}
\mathscr{L} = {} & \ln c^1 + \beta \ln c^2 + \mu_{-1}^c \left[\ln c_0^2 - \ln(c_0^2)^{\bullet}\right] + \phi^y \left[X^{\alpha_1} K^{\alpha_3} - c^1 - c^2 - K\right] + \\
& + \phi_0^y \left[X^{\alpha_1} K_0^{\alpha_3} - c^1 - c_0^2 - K\right] + \phi^R \left[\mathrm{g}(R) - \mathrm{h}(R)X\right] + \\
& + \phi_0^R \left[R_0 + \mathrm{g}(R_0) - \mathrm{h}(R_0)X - R\right],
\end{aligned}$$

yields the following FOCs:

$$\frac{(c^2)^{\bullet}}{\beta (c^1)^{\bullet}} = 1 + \frac{(\phi_0^y)^{\bullet}}{\phi^y)^{\bullet}}, \tag{10.24a}$$

$$\frac{(\mu_{-1}^c)^{\bullet}}{(c_0^2)^{\bullet}} = (\phi_0^y)^{\bullet}, \tag{10.24b}$$

$$\alpha_3 \frac{Y^{\bullet}}{K^{\bullet}} = 1 + \frac{(\phi_0^y)^{\bullet}}{(\phi^y)^{\bullet}}, \tag{10.24c}$$

$$\alpha_1 \left[(\phi_0^y)^{\bullet}) \frac{(Y_0)^{\bullet}}{(X_0)^{\bullet}} + (\phi^y)^{\bullet} \frac{Y^{\bullet}}{X^{\bullet}} \right] = \phi^R \mathrm{h}(R^{\bullet}) + \phi_0^R \mathrm{h}(R_0), \tag{10.24d}$$

$$(\phi^R)^{\bullet} \left[\mathrm{g}'(R^{\bullet}) - \mathrm{h}'(R^{\bullet}) X^{\bullet}\right] = (\phi_0^R)^{\bullet}. \tag{10.24e}$$

Proposition 10.2 characterizes long-run intergenerationally efficient and inefficient solutions.

Proposition 10.2 (Long-run intergenerationally efficient resource stock and resource harvest). *The Golden rule allocation is identified by the unique steady state solution* (R^{MSY}, X^{MSY}) *which satisfies* $g'(R^{MSY}) - h'(R^{MSY})X^{MSY} = 0$ *and hence* $X^{\bullet} = X_{MSY}$ *and* $R^{\bullet} = R_{MSY}$ *(maximum sustainable yield level). A non-trivial steady state characterized by* $R^{\bullet} < R^{MSY}$ *and thus* $g'(R^{\bullet}) - h'(R^{\bullet})X^{\bullet} > 0$ *is dynamically (intergenerationally) efficient.*

Proof. The Golden rule emerges where $(\mu_{-1}^{c})^{\bullet} = 0$ and hence $(\phi_0^y)^{\bullet} = 0$ and $(\phi_0^R)^{\bullet} = 0$. Due to $(\phi_0^R)^{\bullet} = 0$ steady state utility is maximized when $g'(R^{MSY}) = h'(R^{MSY})X^{MSY}$, i.e. the resource stock is at its maximum sustainable yield level, which clearly differs from the maximum sustainable yield level without harvest costs.

When, however, $(\mu_{-1}^{c})^{\bullet} > 0$, it follows that $(\phi_0^y)^{\bullet} > 0$ and hence $(c^2)^{\bullet}/(\beta(c^1)^{\bullet}) = 1 + (\phi_0^y)^{\bullet}/(\phi^y)^{\bullet} > 1$ and therefore we get the marginal productivity of capital $\alpha_3 Y^{\bullet}/K^{\bullet} = 1 + (\phi_0^y)^{\bullet}/(\phi^y)^{\bullet} > 1$. Furthermore, from (10.24e) follows that $g'(R^{\bullet}) - h'(R^{\bullet})X^{\bullet} > 0$ which implies that $R^{\bullet} < R^{MSY}$. □

Proposition 10.3 (Long-run intergenerational (in-)efficiency of steady-state market equilibrium). *Steady-state market equilibria with* $r > 1$ $(r = 1)$ *are long-run intergenerationally efficient (Golden rule) and thus also long-run dynamically efficient. Market equilibria with* $0 < r < 1$ *are dynamically inefficient and thus also intergenerationally inefficient.*

Proof. From (10.9) in the steady state we have $r = 1 + g'(R) - h'(R)X$. Clearly, with $r > 1$ $(r = 1)$ $g'(R) - h'(R)X > 0$ which coincides with the FOC for intergenerational efficiency when $R = R^{\bullet} < R^{MSY}$ $(R^{\bullet} = R^{MSY})$. Otherwise, $R > R^{MSY}$. □

The two possible cases are illustrated by Figs. 10.1– 10.3. In Fig. 10.1, the *XX*-locus cuts the *RR*-locus in the area where the slope of the *RR*-locus is positive—the stationary resource stock and thus, because of the no–arbitrage condition, the man-made capital stock exhibit a positive own rate of return (underaccumulation of the resource stock occurs).[4] This resource stock is below the Golden rule resource stock $R < R^{GR} = R^{MSY}$ which coincides with the maximum sustainable yield level where $g'(R^{MSY}) = h'(R^{MSY})X^{MSY}$ holds.

The second case is illustrated by Figs. 10.2 and 10.3 in which $R > R^{MSY}$ and the *RR*-locus is negatively sloped and which is also compatible with Prop. 10.1, as the proof to Prop. 10.2 shows. This case is, however, intergenerationally inefficient since a central planner could increase welfare of the present and all future generations by a reduction in resource accumulation.

[4] Dynamic efficiency is in line with stylized facts of advanced economies in which the real interest rate is strictly positive $(r > 0)$ and hence also the own rate of return of the resource stock is, due to the no arbitrage condition, positive: $g'(R) - h'(R)X > 0,\ R < R^{MSY}$.

10.6 Stability of Non-Trivial Steady State

For stability of the non-trivial steady state in the present model, the slopes of the XX-locus and the RR-locus at the stationary state are essential. In particular, saddle point stability of the steady state requires that at the steady state the XX-locus cuts the RR-locus from below and hence the sign of the slopes of both loci are derived in Prop. 10.4:

Proposition 10.4 (Slopes of XX and RR loci at the steady state). *Define* $\Omega \equiv \alpha_3 \mathrm{h}'X - (1+\mathrm{g}'-\mathrm{h}'X)\partial \mathrm{A}/\partial R - [\mathrm{g}''-\mathrm{h}''X]\mathrm{A}$ *and* $\Theta \equiv (1+\mathrm{g}'-\mathrm{h}'X)\partial \mathrm{A}/\partial X - \mathrm{h}'\mathrm{A} - \alpha_3 \mathrm{h}$.[5] *Then, the slope of the XX-locus at the non-trivial steady state is given by*

$$\mu'(R) = \frac{\Omega}{\Theta}. \tag{10.25}$$

For the regeneration and harvest cost function specified by (10.1) *and* (10.2), $\Theta > 0$ *and hence $\mu'(R)$ is positive for $\Omega \geq 0$ while it is negative for $\Omega < 0$. The slope of the RR-locus at the non-trivial stationary state, given by*

$$\varphi'(R) = \frac{\mathrm{g}' - \mathrm{h}'X}{\mathrm{h}}, \tag{10.26}$$

is positive for $R \in [0, R^{MSY})$ while it is negative for $R \in (R^{MSY}, R_{max}]$.

Proof. To derive the slopes of the loci at the steady state (10.25)–(10.26), we totally differentiate the intertemporal equilibrium dynamics (10.14) and (10.15) evaluated at the steady state with respect to X and R and drop the cumbersome notation of functions g, h and A. For the specified regeneration and harvest cost functions (10.1) and (10.2), the sign of $\partial \mathrm{A}/\partial X$ is unambiguously positive, but the sign of $\partial \mathrm{A}/\partial R$ is ambiguous.

By definition, at the maximum sustainable yield level R^{MSY} we have $\varphi'(R^{MSY}) = 0$, and hence $\varphi'(R) > 0$ for $R \in [0, R^{MSY})$ while $\varphi'(R) < 0$ for $R \in [R^{MSY}, R_{max})$. To sign $\mu'(R)$, it is straightforward to verify that $\Theta > 0$ for the specified regeneration and harvest cost functions, (10.1) and (10.2). Hence $\mu'(R) > 0 \Longleftrightarrow \Omega > 0$ and $\mu'(R) < 0 \Longleftrightarrow \Omega < 0$. □

As is typical for logistic regeneration, the slope of the RR locus, $\varphi'(R_t)$, is increasing for low values of R while it is decreasing after the maximum sustainable yield level R^{MSY}. The slope of the XX-locus, $\mu(R_t)$, is positive at the origin (see Proof to Prop. 10.1) but generally ambiguous over the whole range of R_t (Θ is certainly positive but Ω involves both positive and negative terms). Figures 10.1– 10.3, known as phase diagrams, illustrate the three possible cases of slopes of the loci at the steady state solution: either both loci are positively sloped at the steady state (Fig. 10.1), or the RR-locus is negatively sloped and the XX-locus is positively sloped (Fig. 10.2), or both loci are negatively sloped (Fig. 10.3).[6]

[5] Clearly, Ω in this chapter has a different interpretation than in the previous chapters.

[6] The graphical illustrations require a numerical specification of the model. The parameters are chosen, in accordance with Props. 10.1– 10.5 as $\beta = 0.9, \lambda = 30, \rho = 1, R_{\max} = 500$, and for

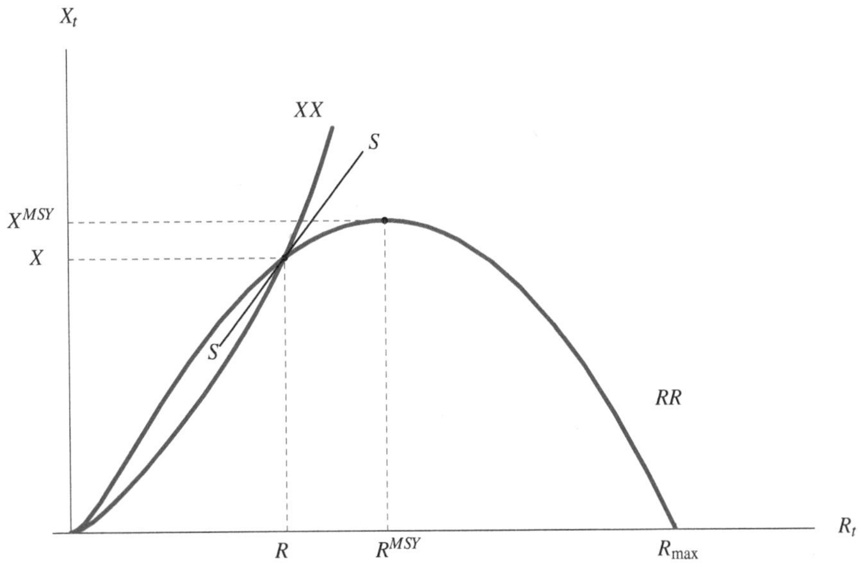

Fig. 10.1 Positively sloped RR and XX loci ($\varphi'(R) > 0$, $\mu'(R) > 0$) at an intergenerationally efficient steady state ($R < R^{MSY}$)

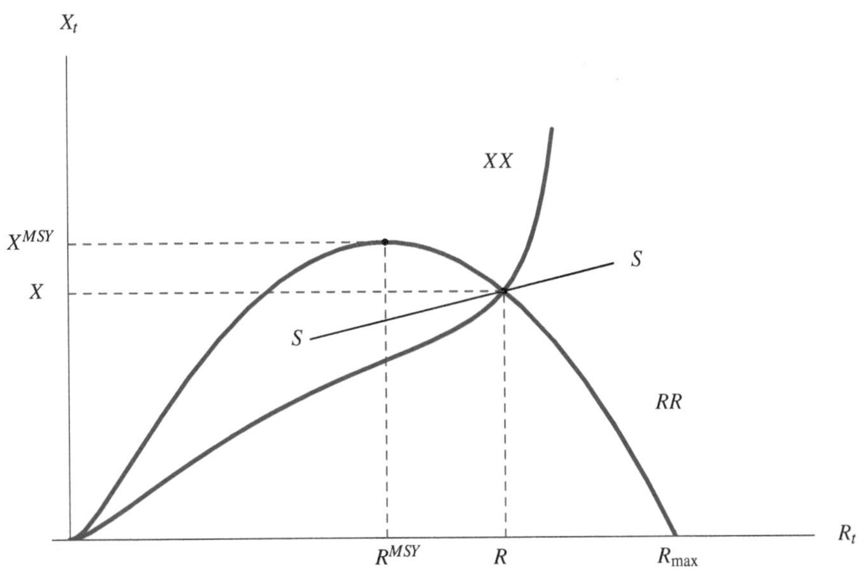

Fig. 10.2 Negatively sloped RR-locus ($\varphi'(R) < 0$) and positively sloped XX-locus ($\mu'(R) > 0$) at an intergenerationally inefficient steady state ($R > R^{MSY}$)

Fig. 10.1 $\alpha_1 = 0.05, \alpha_2 = 0.65, \alpha_3 = 0.3, \pi = 0.86$, for Fig. 10.2 $\alpha_1 = 0.1, \alpha_2 = 0.8, \alpha_3 = 0.1, \pi = 1.2$, and for Fig. 10.3 $\alpha_1 = 0.195, \alpha_2 = 0.8, \alpha_3 = 0.005, \pi = 1.25$.

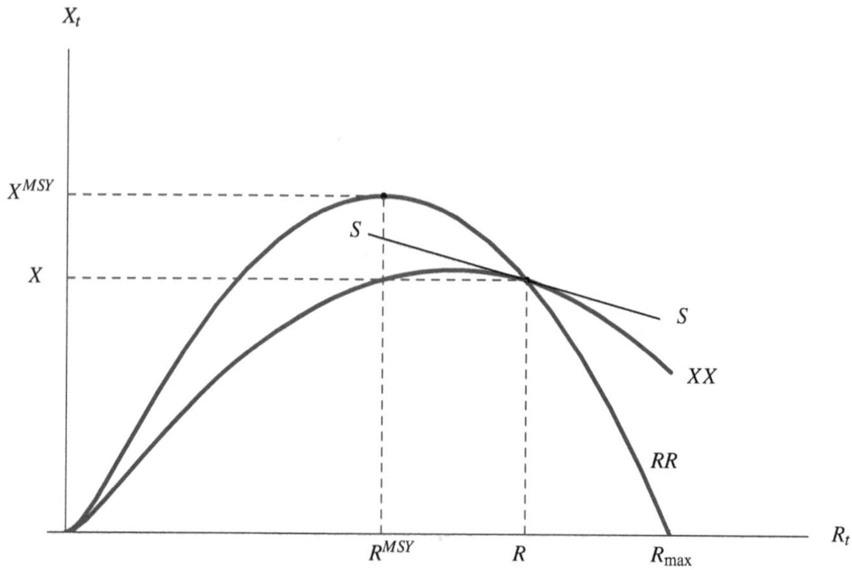

Fig. 10.3 Negatively sloped RR and XX ($\varphi'(R) < 0$, $\mu'(R) > 0$) loci at a intergenerationally inefficient steady state ($R > R^{MSY}$)

The next step comprises deriving the elements of the Jacobian, which follow from partially differentiating the system of equations (10.14) and (10.15) and acknowledging that $A = \alpha_3 \mathrm{h} X/(1+\mathrm{g}'-\mathrm{h}'X)$ according to (10.19):

$$J_{11} = \frac{\partial X_{t+1}}{\partial X_t} = \frac{(1+\mathrm{g}'-\mathrm{h}'X)\{\alpha_3 \mathrm{h}'X\mathrm{h} - \mathrm{h}\mathrm{A}[\mathrm{g}''-\mathrm{h}''X] + (1+\mathrm{g}'-\mathrm{h}'X)\,\partial \mathrm{A}/\partial X\}}{\alpha_3\,(1+\mathrm{g}')\,\mathrm{h}}, \tag{10.27a}$$

$$J_{12} = \frac{\partial X_{t+1}}{\partial R_t} = \frac{(1+\mathrm{g}'-\mathrm{h}'X)^2\{\partial \mathrm{A}/\partial R + \mathrm{A}[\mathrm{g}''-\mathrm{h}''X] - \alpha_3 \mathrm{h}'X\}}{\alpha_3\,(1+\mathrm{g}')\,\mathrm{h}}, \tag{10.27b}$$

$$J_{21} = \frac{\partial R_{t+1}}{\partial X_t} = -\mathrm{h} < 0, \tag{10.27c}$$

$$J_{22} = \frac{\partial R_{t+1}}{\partial R_t} = 1+\mathrm{g}'-\mathrm{h}'X > 0. \tag{10.27d}$$

With this information at hand, Prop. 10.5 states that the steady state is a saddle regardless whether the steady state is intergenerationally efficient or not and regardless whether the loci are positively or negatively sloped at the steady state.

Proposition 10.5 (Stability of the non-trivial stationary state). *Let the regeneration and harvest cost function by specified by* (10.1) *and* (10.2) *and let the slope conditions in Prop. 10.1 hold. Then, the non-trivial stationary state represents a non–oscillating stable saddle point.*

Proof. We start by signing the elements of the Jacobian. It is straightforward to verify that the denominator of J_{11} and J_{12} is positive due to (10.19). According to the proof of Prop. 10.4, for the specified regeneration and harvest cost functions the denominator of (10.21) is negative: $\partial \text{LHS}/\partial X - \partial \text{RHS}/\partial X < 0$. Acknowledging that, according to (10.19), $\alpha_3 \text{h} + \text{A}\text{h}' = (1+\text{g}')\text{A}/X > 0$ holds at the steady state, it follows that the numerator of J_{11} is certainly positive and hence $J_{11} > 0$. Regarding the sign of the numerator of J_{12}, either $\partial \text{RHS}/\partial R - \partial \text{LHS}/\partial R \leq 0$ and hence $J_{12} \leq 0$ (case 1) or the opposite holds: $\partial \text{RHS}/\partial R - \partial \text{LHS}/\partial R > 0$ and hence $J_{12} > 0$ (case 2).

Then, the eigenvalues of the Jacobian at the non-trivial stationary state are real and distinct if the discriminant of the Jacobian, $\triangle J$ is positive which is certainly true in case 1 since

$$\triangle J = \left[J_{11} - (1+\text{g}' - \text{h}'X)\right]^2 - 4hJ_{12} > 0.$$

In case 2, the sign of $\triangle J$ seems to be ambiguous. However, for all admissible parameter combinations which imply $J_{12} > 0$ it can be verified numerically that $\triangle J > 0$, too. Thus, $\triangle J > 0$ regardless whether $J_{12} \gtrless 0$.

For the eigenvalues to be strictly positive, the trace has to be larger than one and the determinant at (R, X) has to be positive:

$$\text{tr}J = J_{11} + J_{22} > 1,$$

$$\det J = \left\{(1+\text{g}' - \text{h}'X)\frac{\partial \text{A}}{\partial X} + \text{h}\frac{\partial \text{A}}{\partial R}\right\} \frac{(1+\text{g}' - \text{h}'X)^2}{\alpha_3(1+\text{g}')\text{h}}.$$

If $(1+\text{g}' - \text{h}'X)\partial \text{A}/\partial X + \text{h}\partial \text{A}/\partial R > 0$, the determinant is positive. But this requirement is fulfilled, due to functional specification for the harvest and regeneration functions; in that case, this expression reduces to

$$-\frac{\lambda \text{g}\left(R^2 - \rho(\rho+\lambda)\right)}{(\rho+R+\lambda)^2(\rho+R)^2} + \frac{\lambda R(1+\text{g}')}{(\rho+R+\lambda)(\rho+R)} < (1+\text{g}')\frac{\alpha_2}{\alpha_1} - \text{g}''R,$$

where on the left hand side the first term is certainly negative, and the second term is considerably smaller than the first term on the right hand side (since $\lambda R\left[(\rho+R+\lambda)(\rho+R)\right]^{-1} < 1$ and $\alpha_2 > \alpha_1$), and thus $\det J > 0$.
Finally, to guarantee that the first of the eigenvalues, ψ_1, is larger than one, and the second, ψ_2, is between zero and one, it is required that $1 - \text{tr}J + \det J < 0$:

$$1 - \text{tr}J + \det J = \frac{(\text{g}' - \text{h}'X)\Theta - \text{h}\Omega}{\alpha_3 \text{h} + \text{h}'\text{A}} = \frac{1}{\text{h}\Theta}\frac{\varphi'(R) - \mu'(R)}{\alpha_3 \text{h} + \text{h}'\text{A}}.$$

To sign $1 - \text{tr}J + \det J$, consider first that the stationary state is intergenerationally efficient, $R \leq R^{MSY}$. Because of Prop. 10.1, $\mu'(R) > \varphi'(R) \geq 0$ for $R \in (0, R^{MSY}]$ and hence $1 - \text{tr}J + \det J < 0$. For a dynamically inefficient steady state, $R \in (R^{MSY}, R_{max}]$, $\varphi'(R) < 0$ and $\mu'(R) \gtrless 0$, depending on the sign of Ω. But the slope condition in Prop. 10.1 still requires that $\mu'(R) > \varphi'(R)$, and hence $1 - \text{tr}J + \det J < 0$, too. Thus, the non-trivial stationary state is a non–oscillating saddle. □

The gist of Prop. 10.5 is that the existence condition ensures saddle–point stability of the non-trivial steady state. This confirms that in OLG models with log–linear preferences and Cobb–Douglas technology existence, uniqueness and stability of steady states are closely connected (see in another model context Ono, 2002; Farmer, 2006). As shown in the proof, saddle point stability is equivalent to the fact that the XX-locus cuts the RR-locus from below at the steady state. And this is the case since near the origin the XX-locus is below the RR-locus while at high resource stock levels (near the carrying capacity) the XX-locus lies above the RR-locus.

However, while the existence condition is also pivotal for dynamic stability in the no–harvest cost case, as shown in Prop. 9.3, or in the log–linear utility case but without man-made capital (Koskela et al, 2002, Prop. 3), for the present model with man-made capital and harvest costs this property is much harder to demonstrate: since positive and negative terms prevail in the numerator of J_{12} sign ambiguities results for the slope of the XX-locus as well as for the determinant of the Jacobian, which vanish when setting harvest cost to zero (and hence $\text{h}(R) = 1$ and $\text{h}'(R) = 0$).

10.7 Uniqueness of Non-Trivial Steady State

Since Krutilla and Reuveny (2004) conclude that harvest costs can lead to multiple solutions in the ILA context, it remains to be investigated whether harvest costs can lead to multiple solutions in an OLG framework, too. In principle the XX-locus could coincide with the RR-locus more than once particularly for small values of R where the impact of harvest costs on resource dynamics is most severe (see the initially slightly convex shape of the RR-locus). However, it follows from Props. 10.1 and 10.4 that the non-trivial steady state is unique also in our model with physical harvest cost:

Corollary 10.1 (Uniqueness of non-trivial stationary state). *Let the regeneration and harvest cost function be specified by* (10.1) *and* (10.2)*. Then, it follows from Props. 10.1 and 10.4 that the non-trivial steady state solution* (R, X) *is unique.*

Proof. For the (local) uniqueness of the non-trivial steady state solution, the monotonicity of both the XX and the RR loci in the neighborhood of the steady state is needed. We distinguish intergenerationally efficient (including the Golden Rule) from intergenerationally inefficient steady state solutions. For intergenerationally efficient steady state solutions (i.e., $R < R^{MSY}$), the stability condition $1 - \text{tr}J + \det J < 0$ implies that $\Omega > 0$ and hence both the RR and XX-locus are monotonically increasing in the neighborhood of (R, X). For dynamic inefficiency (i.e., $R > R^{MSY}$), two cases are to be distinguished: either $\Omega \geq 0$ or $\Omega < 0$. If $\Omega \geq 0$, the RR-locus is monotonically decreasing while the XX-locus is monotonically increasing in the neighborhood of (R,X). If $\Omega < 0$, both the RR-locus and the the XX-locus are monotonically decreasing in the neighborhood of (R,X) but the former decreases more than the latter. Thus, the steady state is locally unique, regardless whether it is intergenerationally efficient or not. □

Geometrically considered, Corollary 10.1 requires that both the RR and the XX-locus in the neighborhood of the steady state are monotonic and that hence the RR and XX-locus intersect locally once only. But this is ensured by the existence condition in Prop. 10.1 which requires that the slope of the RR-locus near the origin is larger than that of the XX-locus.

Knowing that the steady state solution is a unique saddle point, we approximate linearly the equilibrium dynamics at the non-trivial steady state, i.e. the saddle path or stable arm, by using the elements and the second eigenvalue of the Jacobian.

Corollary 10.2 (Linearized dynamics around steady state). *Let the second eigenvalue $0 < \psi_2 < 1$ be given from the proof to Prop. 10.5. A linear approximation of the equilibrium dynamics of R_t, X_t, and K_t evaluated at the saddle point (R,X) takes the following form:*

$$X_t = X + \frac{(1+\mathrm{g}'-\mathrm{h}'X)-\psi_2}{\mathrm{h}}(R_t - R), \tag{10.28a}$$

$$R_{t+1} = R(1-\psi_2) + \psi_2 R_t, \; given\, R_0 > 0, \tag{10.28b}$$

$$K_{t+1} = \alpha_3 \frac{X_{t+1}}{1+\mathrm{g}'(R_{t+1}) - \mathrm{h}'(R_{t+1})X_{t+1}} \frac{\mathrm{h}(R_{t+1})}{\mathrm{h}(R_t)} (X_t)^{\alpha_1 - 1} (K_t)^{\alpha_3}, \; given\, K_0 > 0. \tag{10.28c}$$

As a consequence of Prop. 10.5, for $\Omega \geq 0$ and $R \in (0, R_{max}]$, $(1+\mathrm{g}'-\mathrm{h}'X)-\psi_2 > 0$ and hence the saddle path (labeled by S) at (R,X) is positively sloped. For $\Omega < 0$ and consequently $R > R^{MSY}$, $(1+\mathrm{g}'-\mathrm{h}'X)-\psi_2 < 0$ and hence the saddle path at (R,X) is negatively sloped.

Proof. The equilibrium dynamics at the non-trivial steady state are approximated linearly by:

$$\begin{pmatrix} X_{t+1} \\ R_{t+1} \end{pmatrix} = J(X,R) \begin{pmatrix} X_t \\ R_t \end{pmatrix} + (I - J(X,R)) \begin{pmatrix} X \\ R \end{pmatrix}, \tag{10.29}$$

where (X,R) denote the steady state values, $J(X,R)$ is the Jacobian evaluated at the stationary state and I denotes the unitary matrix. After deriving the general solution to the system of difference equations and normalizing the eigenvector, we gain the linearized dynamics of the resource stock and resource harvest, and the dynamics of man-made capital follow from (10.16). □

The two cases of Prop. 10.2 are illustrated by Figs. 10.1– 10.3, either the saddle path S in the neighborhood of the steady state is positively sloped (Fig. 10.1 and 10.2) or negatively sloped (Fig. 10.3).

10.8 Conclusions

In this chapter a physical harvest cost function was introduced in an OLG model with a renewable natural resource and with man-made capital. We proved the existence, uniqueness, and stability of a strictly positive (non-trivial) steady state solution. It turns out that while the notion of physical harvest costs complicates the analysis considerably compared to the no–harvest cost case (in particular in regard to the existence and uniqueness of the XX-locus), the general properties that uniqueness and stability are a consequence from the existence condition carry over from a comparable model without harvest cost.[7] If, moreover, the steady-state resource stock is smaller than the maximum sustainable yield level, the solution is dynamically (intergenerationally) efficient.

In this chapter we neglected shocks to the parameters of the harvest cost function and the regeneration function. The impacts of these shocks on the intertemporal market equilibrium solution will be considered in the following chapter.

Appendix A: Individual Optimization and Market Clearing Conditions with Endogenous Labor Supply

In this section we briefly outline the changes when, instead of physical harvest costs entering the dynamics of the resource stock, resource harvest has to compete with commodity production for labor. As in partial equilibrium resource harvest models, we assume now that harvest requires effort (labor). Thus, the young household splits total labor supply (which will again, without loss of generality, be normalized to 1) between resource harvest (E_t for effort) and labor in the production of the commodity ($L_t^y = 1 - E_t$). This effort is the product of harvest costs h and units harvested X.

To account for such a division of labor between the two alternative purposes, the first period budget constraint and the accumulation equation of the natural resource stock take the following form instead of (10.5) and (10.3) above:

$$p_t\, R_t + c_t^1 + K_{t+1} = w_t\left[1 - \mathrm{h}(R_t)X_t\right] + q_t X_t, \tag{10.30}$$

$$R_{t+1} = R_t + \mathrm{g}(R_t) - X_t, \tag{10.31}$$

where $\mathrm{h}(R_t)$ denotes now the stock-dependent harvest costs which can be specified as:

$$\mathrm{h}(R_t) = \frac{\lambda}{\rho + R_t}, \tag{10.32}$$

[7] One obvious objection against this point is that this result depends on the functional specification of the harvest and resource growth function by (10.1) and (10.2). While this is true, the functional forms are quite popular in resource models, as argued in Sect. 10.2.

with the harvest cost parameters $\lambda > 0$ and ρ small (but positive). Accordingly, the scarcer the resource the more costly it is to harvest and λ reflects the effort or difficulty necessary to harvest one unit. For a fishery, the value of λ can be understood as the time units necessary to catch one fish while for a forest it represents the time units needed to harvest one tree.

The household's FOCs under endogenous labor supply read as follows:

$$\frac{c_{t+1}^2}{\beta c_t^1} = r_{t+1}, \tag{10.33}$$

$$q_t = \frac{p_{t+1}}{r_{t+1}} + w_t \mathrm{h}(R_t), \tag{10.34}$$

$$p_t = \frac{p_{t+1}}{r_{t+1}}\left[1 + \mathrm{g}'(R_t)\right] - w_t \frac{\mathrm{h}'(R_t)}{\mathrm{h}(R_t)} X_t. \tag{10.35}$$

Finally, labor market clearing requires now $N_t^y = 1 - \mathrm{h}(R_t)X_t$ and commodity market clearings changes to:

$$(X_t)^{\alpha_1}(1 - \mathrm{h}(R_t)X_t)^{\alpha_2}(K_t)^{\alpha_3} = c_t^1 + c_t^2 + K_{t+1}. \tag{10.36}$$

Appendix B: Proof to Lemma 10.1

Restating (10.14) at the XX-locus, i.e. evaluated for $X_{t+1} = X_t, \forall t$, yields

$$\underbrace{\alpha_3 h(R_{t+1})X_t}_{\mathrm{LHS}(R_t,X_t)} = \underbrace{\left(1 + \mathrm{g}'(R_{t+1}) - h'(R_{t+1})X_t\right)\mathrm{A}(R_t,X_t)}_{\mathrm{RHS}(R_t,X_t)}. \tag{10.14'}$$

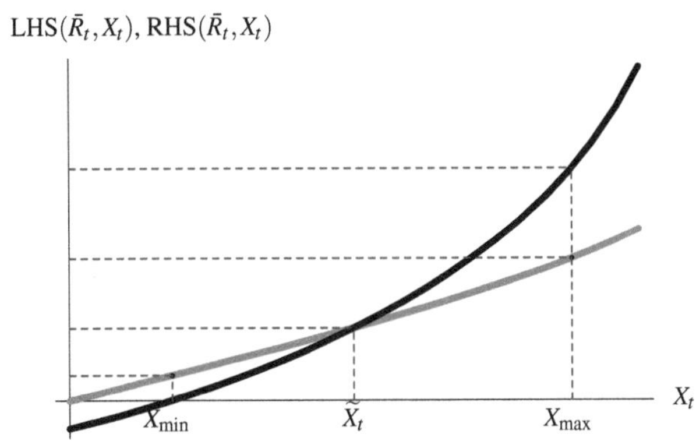

Fig. 10.4 Condition for the XX-locus to exist for given R_t

To show that (10.14') has at least one solution $(R_t, X_t) \in [0, R_{max}] \times (X_{min}(R_t), X_{max}(R_t))$, and that therefore the XX-locus exists, we first show that LHS $(R_t, X_t) -$ RHS (R_t, X_t) contains an arbitrary $\bar{R}_t \in (0, R_{max}]$ for which LHS$(\bar{R}_t, X_t)$ $-$RHS$(\bar{R}_t, X_t) = 0$ (see also Fig. 10.4 where functions LHS(R_t, X_t) (gray graph) and RHS(R_t, X_t) (black graph) are depicted for a particular value of $\bar{R}_t = R_t \in (0, R_{\max}]$). For that purpose, we evaluate LHS $(R_t, X_t) -$ RHS (R_t, X_t) at $X_t = X_{\min}(R_t)$ and $X_t = X_{\max}(R_t)$. At $X_{\min}$, LHS $(X_{\min}(R_t)) = \alpha_3 \mathrm{h}(R_{t+1}) X_{\min}(R_t) > 0$ and, since the definition of $X_{\min}$ implies that $\mathrm{A}(R_t, X_{\min}(R_t)) = 0$, it follows that RHS $(X_{\min}(R_t)) = 0$ and LHS $(X_{\min}(R_t)) >$ RHS $(X_{\min}(R_t))$.

For $X_{\max}(R_t) \Longleftrightarrow R_{t+1} = 0$, we have LHS $(R_t, X_{\max}(R_t)) <$ RHS $(R_t, X_{\max}(R_t))$ for all $R_t \in (0, R_{\max}]$ if the following condition holds:

$$\alpha_3 \mathrm{h}(0) < \left(1 + \mathrm{g}'(0) - \mathrm{h}'(0) X_{\max}(R_t)\right) \left[(\alpha_1 + \alpha_2 \sigma) \mathrm{h}(R_t) - \alpha_1 \xi(R_t)\right], \tag{10.37}$$

where $\xi(R_t) \equiv \dfrac{\gamma \phi(R_t) + 1 + \mathrm{g}'(R_t) - \sigma \mathrm{h}'(R_t)(R_t + \mathrm{g}(R_t))}{1 + \mathrm{g}'(R_t)/R_t}$. Acknowledging that $g(R_t) \equiv r\left(R_t - R_t^2/R_{\max}\right)$ and $h(R_t) \equiv 1 + \lambda/(\rho + R_t)$, we have $\xi(R_t) = \xi_0(R_t) + \xi_1(R_t)$ with $\xi_0(R_t) \equiv \frac{(1+r)R_{max} - (2-\gamma) r R_t}{(1+r)R_{max} - r R_t}$ and $\xi_1(R_t) \equiv \frac{\sigma \lambda R_t}{(\rho + R_t)^2}$. Since $\xi_0(R_t)$ is a nearly constant, slightly positively sloped function and $\xi_1(R_t)$ is strictly concave, there is a unique maximum of $\xi'(R_t)$ at $\xi'(R_t) = 0$. It is not difficult to see that $\xi'(R_t) = 0$ at $R_t \approx \rho$. Hence, for ρ near zero, $\xi(0) = 1$ maximizes $\xi(R_t)$ and therefore for small ρ, condition (10.37) can be restated as:

$$\alpha_3 \mathrm{h}(0) < (1 + \mathrm{g}(0)) \left((\alpha_1 + \alpha_2 \sigma) \mathrm{h}(0) - \alpha_1\right). \tag{10.17}$$

Since, moreover, LHS(R_t, X_t) and RHS(R_t, X_t) are continuous on $(X_{min}(R_t), X_{max}(R_t))$ and positively sloped functions according to

$$\frac{\partial \mathrm{LHS}(R_t, X_t)}{\partial X_t} = -\alpha_3 \mathrm{h}'(R_{t+1}) \mathrm{h}(R_t) X_t + \alpha_3 \mathrm{h}(R_{t+1}) > 0, \tag{10.38}$$

$$\begin{aligned} \frac{\partial \mathrm{RHS}(R_t, X_t)}{\partial X_t} &= \left(1 + \mathrm{g}'(R_{t+1}) - \mathrm{h}'(R_{t+1}) X_t\right) \frac{\partial \mathrm{A}(R_t, X_t)}{\partial X_t} - \mathrm{A}(R_t, X_t) \times \\ &\times \left[\mathrm{g}''(R_{t+1}) \mathrm{h}(R_t) + \mathrm{h}'(R_{t+1}) - \mathrm{h}''(R_{t+1}) \mathrm{h}(R_t) X_t\right] > 0, \end{aligned} \tag{10.39}$$

with $\partial \mathrm{A}(R_t, X_t)/\partial X_t = (\alpha_1 + \alpha_2 \sigma) \mathrm{h}(R_t) + \alpha_1 \sigma \mathrm{h}'(R_t) R_t$, the Intermediate Value Theorem ensures that there exists at least one X_t such that LHS $(R_t, X_t) -$ RHS (R_t, X_t) $= 0$, for all $R_t \in (0, R_{max}]$. For $R_t = 0$, we have $A(R_t, X_t) = 0$ and $X_t = 0$ and therefore there exists at least one X_t for all $R_t \in [0, R_{max}]$ for which LHS $(R_t, X_t) -$ RHS $(R_t, X_t) = 0$ holds.

Knowing that for a given $R_t \in [0, R_{max}]$ a solution to LHS $(R_t, X_t) -$ RHS $(R_t, X_t) =$ 0 exists and ∂LHS$/\partial X_t$ $-\partial$RHS$/\partial X_t \neq 0$, the Implicit Function Theorem assures that there exists a single valued, continuously differentiable function $X_t = \mu(R_t)$. □

References

Armstrong CW, Sumaila UR (2001) Optimal allocation of TAC and the implications of implementing an ITQ management system for the North-East arctic cod. Land Economics 77(3):350–359

Berck P (1981) Optimal management of renewable resources with growing demand and stock externalities. Journal of Environmental Economics and Management 8(2):105–117

Brown GM (2000) Renewable natural resource management and use without markets. Journal of Economic Literature 38(4):875–914

Clark CW (1990) Mathematical Bioeconomics: The Optimal Management of Renewable Resources, 2nd edn. Wiley, NY

Elíasson L, Turnovsky SJ (2004) Renewable resources in an endogenously growing economy: balanced growth and transitional dynamics. Journal of Environmental Economics and Management 48:1018–1049

Escapa M, Prellezo R (2003) Fishing technology and optimal distribution of harvest rates. Environmental and Resource Economics 425:377–394

Farmer K (2006) Reducing public debt under dynamic efficiency: Transitional dynamics in Diamond's OLG model. Atlantic Economic Journal 34:195–208

Galor O (1992) A two–sector overlapping–generations model: A global characterization of the dynamical system. Econometrica 60(6):1351–1386

Galor O, Ryder HE (1989) Existence, uniqueness, and stability of equilibrium in an overlapping–generations model with productive capital. Journal of Economic Theory 49:360–375

Hayashi F (1982) Tobin's margin q and average q: A neoclassical interpretation. Econometrica 50:213–24

Heaps T, Neher PA (1979) The economics of forestry when the rate of harvest is constrained. Journal of Environmental Economics and Management 6:297–319

Koskela E, Ollikainen M, Puhakka M (2002) Renewable resources in an overlapping generations economy without capital. Journal of Environmental Economics and Management 43(3):497–517

Krutilla K, Reuveny R (2004) A renewable resource–based Ramsey model with costly resource extraction. Environmental and Resource Economics 27:165–185

Neher P (1990) Natural Resource Economics: Conservation and exploitation. Cambridge University Press, Cambridge, New York

Ono T (2002) The effects of emission permits on growth and the environment. Environmental and Resource Economics 21:75–87

Rhee C (1991) Dynamic efficiency in an economy with land. Review of Economic Studies 58

Schaefer MB (1954) Some aspects of the dynamics of populations important to the management of commercial marine fisheries. Bulletin of Inter–American Tropical Tuna Commission 1:25–56

Shone R (1997) Economic Dynamics. Phase diagrams and their economic application. Cambridge University Press, Cambridge, New York

Smith VL (1968) Economics of production from natural resources. American Economic Review 58:409–432

Tahvonen O, Kuuluvainen J (2000) The economics of natural resource utilization. In: Folmer H, Gabel HL (eds) Principles of Environmental and Resource Economics, 2nd edn, Edward Elgar, Chelthenham, pp 665–699

Chapter 11
Effects of Harvest Cost and Biological Shocks

11.1 Introduction

Until now we did not consider changes in the basic parameters of our log-linear CD OLG model with a renewable natural resource. Parameter shocks both to the economic system and shifts in the natural resource influence the dynamics of the market system. In this chapter, we investigate two forms of such shocks: shocks to the regeneration ability of natural resources and a push in harvest costs. Potential sources of the former shocks are flooding, landslide or windthrow, infectious diseases or invasive species that displace native ones. Both economic factors (technological change, opportunity costs of harvesting etc.) and natural ones (remoteness, weather and climatic conditions etc.) can cause a push in harvest costs. We want to investigate the economic impacts of these two different types of resource shocks in the steady state and during the transition phase.

Contrary to first thoughts, that the nature of the resource technology shock is irrelevant for the characteristics of the harvest dynamics, the analysis of the transitional dynamics indicates that the harvest cost shock generates harvest transition dynamics arising qualitatively different from the dynamics in response to shocks to the parameters in the natural growth function. Moreover, a shock in resource technologies causes a revaluation of the resource stock and of the capital stock over time and thereby effects the different generations in different ways.

We start by an analysis of the steady-state effects of a harvest cost and of a biological shock. Building on the steady-state results, the transitional dynamics along the saddle path following the shocks are analyzed and compared across types of shocks. The final section discusses the results and provides suggestions for extensions of the analysis.

K. Farmer and B. Bednar-Friedl, *Intertemporal Resource Economics: An Introduction to the Overlapping Generations Approach*, DOI 10.1007/978-3-642-13229-2_11,

11.2 Steady-State Effects of Cost and Biological Shocks

This section is devoted to the analysis of the effects of two forms of shocks to the natural resource technology on the stationary–state values of R and X (steady-state analysis of resource technology shocks). This analysis requires a functional specification of the natural resource growth and the harvest cost function. For the regeneration function, we use as before the logistic function,

$$g(R_t) \equiv \pi \left(R_t - \frac{R_t^{\,2}}{R_{\max}} \right), \tag{11.1}$$

where π denotes the natural growth rate, or increase in units of biomass, and $R_{\max}$ the carrying capacity. The harvest cost function is specified by

$$h(R_t) = 1 + \frac{\lambda}{\rho + R_t}, \tag{11.2}$$

with the harvest cost parameters $\lambda > 0$ and ρ small (but positive). For simplicity, we set $\rho = 1$.

A harvest cost shock is then specified as an exogenous change in the harvest cost parameter λ, while a biological shock will be specified as an exogenous change in the natural growth rate π. To facilitate intuition for these shocks, an increase in harvest cost can be thought of as increasing the effort necessary to harvest one unit of the resource, caused e.g. by the occurrence of flooding, windthrow or landslide. On the other hand, a diminished natural growth rate (i.e., smaller fish, less log biomass) could be the result of changed climatic conditions or an alteration of the habitat or ecosystem where the resource stock is based.

In the following, we first derive the steady-state (long-run) effects of these shocks as a prerequisite for understanding the transition from the pre-shock to the post-shock steady state triggered by these two forms of shocks. Furthermore, we restrict our analysis to the case that the steady state is to the left of the maximum sustainable yield level, i.e. the slope of the RR-locus is positive: $g' - h'X > 0$.[1]

11.2.1 A Shock in Harvest Technologies: The Harvest Cost Parameter λ Rises

Let us first investigate the steady-state impacts of a cost push in the harvest cost parameter, which could emerge due to an increase in the costs associated with the effort necessary to harvest one unit of the resource stock. For that purpose, we acknowledge that the harvest cost function depends on the harvest cost parameter, too: $h(R,\lambda)$. Thus, from now on its partial derivatives will be denoted as $\partial h(R,\lambda)/\partial R \equiv h_R$ (instead of h') and $\partial h(R,\lambda)/\partial \lambda \equiv h_\lambda$ etc.

[1] For clarity of exposition, from now on we drop the cumbersome notation of functions g, h and A.

Recall the intertemporal equilibrium dynamics (10.14) and (10.15) and investigate them at the steady state, i.e. $X = \Phi^X(R,X)$ and $R = \Phi^R(R,X)$. Totally differentiating $X = \Phi^X(R,X)$ and $R = \Phi^R(R,X)$ with respect to X, R and λ yields:

$$\begin{pmatrix} \mathrm{h} & -\Upsilon \\ -\Theta & \Omega \end{pmatrix} \begin{pmatrix} \mathrm{d}X \\ \mathrm{d}R \end{pmatrix} = \begin{pmatrix} -\mathrm{h}_\lambda X \\ \Lambda \end{pmatrix} \mathrm{d}\lambda, \tag{11.3}$$

where

$$\begin{aligned}
\Upsilon &\equiv \mathrm{g}_R - \mathrm{h}_R X, \\
\Omega &\equiv -\left[1 + \mathrm{g}_R - \mathrm{h}_R X\right] \frac{\partial \mathrm{A}}{\partial R} - \left[\mathrm{g}_{RR} - \mathrm{h}_{RR} X\right] \mathrm{A} + \alpha_3 \mathrm{h}_R X > 0, \\
\Theta &\equiv \left[1 + \mathrm{g}_R - \mathrm{h}_R X\right] \frac{\partial \mathrm{A}}{\partial X} - \mathrm{h}_R \mathrm{A} - \alpha_3 \mathrm{h} > 0, \\
\Lambda &\equiv \left(1 + \mathrm{g}_R - \mathrm{h}_R X\right) \frac{\partial \mathrm{A}}{\partial \lambda} - \mathrm{h}_{R\lambda} X \mathrm{A} - \alpha_3 \mathrm{h}_\lambda X,
\end{aligned}$$

and $\partial \mathrm{A}/\partial \lambda = (\alpha_1 + \alpha_2 \sigma) \mathrm{h}_\lambda X + \alpha_1 \sigma \mathrm{h}_{R\lambda} X R > 0$. Then, the steady-state derivatives in (11.3) are given by

$$\frac{\mathrm{d}R}{\mathrm{d}\lambda} = \frac{\left[\mathrm{h}\Lambda - \mathrm{h}_\lambda X \Theta\right]}{\Delta}, \tag{11.4}$$

$$\frac{\mathrm{d}X}{\mathrm{d}\lambda} = \frac{\left[-\mathrm{h}_\lambda X \Omega + (\mathrm{g}_R - \mathrm{h}_R X)\Lambda\right]}{\Delta}, \tag{11.5}$$

where Δ denotes the determinant of the Jacobian J in (11.3), $\Delta = \mathrm{h}\Omega - (\mathrm{g}_R - \mathrm{h}_R X)\Theta$. Prop. 11.1 signs the steady-state effects of a harvest cost push:

Proposition 11.1 (Steady state effects of a harvest cost push). *The steady-state effects of a harvest cost push on the resource stock and on the resource harvest, respectively, are given in general as follows:*

$$\frac{\mathrm{d}R}{\mathrm{d}\lambda} \gtrless 0 \Longleftrightarrow \mathrm{h}\Lambda \gtrless \mathrm{h}_\lambda X \Theta \quad \wedge \quad \frac{\mathrm{d}X}{\mathrm{d}\lambda} \lesseqgtr 0 \Longleftrightarrow \left.\frac{\mathrm{d}X}{\underset{\cdot}{R}}\right|_{XX} \left.\frac{\partial R}{\partial \lambda}\right|_{RR} + \left.\frac{\partial X}{\partial \lambda}\right|_{XX} \lesseqgtr 0,$$

where $\partial X/\partial \lambda_{|XX} \equiv -\Lambda/\Theta$ *and* $\partial R/\partial \lambda_{|RR} \equiv \mathrm{h}_\lambda X/(\mathrm{g}_R - \mathrm{h}_R X)$*).*

Proof. Due to the correspondence principle according to Samuelson (1947) (see, e.g. Gandolfo, 1997, 314), determinant Δ is positive since the stability condition holds, i.e. $1 - tr(J) + det(J) < 0$, see the proof to Prop. 10.5. Hence, $\mathrm{d}R/\mathrm{d}\lambda \gtrless 0$, if and only if $[\mathrm{h}\Lambda - \mathrm{h}_\lambda X \Theta] \gtrless 0$.

After substituting for the slopes of the isoclines ($\mathrm{d}X/\mathrm{d}R_{|RR}$, $\mathrm{d}X/\mathrm{d}R_{|XX}$) and for the partial derivatives ($\partial X/\partial \lambda_{|XX}$ and $\partial R/\partial \lambda_{|RR}$), the sign of $dX/d\lambda$ can be inferred more easily:

$$\frac{\mathrm{d}X}{\mathrm{d}\lambda} = -\mathrm{h}\frac{\Theta}{\Delta} \left.\frac{\mathrm{d}X}{\mathrm{d}R}\right|_{RR} \left[\left.\frac{\mathrm{d}X}{\mathrm{d}R}\right|_{XX} \left.\frac{\partial R}{\partial \lambda}\right|_{RR} + \left.\frac{\partial X}{\partial \lambda}\right|_{XX} \right].$$

Since $dX/dR_{|RR} > 0$, $\Theta > 0$, $\Delta > 0$ and $h > 0$, the sign of $dX/d\lambda$ is negative if the term in square brackets is positive. □

The ambiguity of the sign of the numerator of (11.4) confirms the general insight of Gandolfo (1997, 309) that the correspondence principle helps to sign the denominator in the steady-state analysis but not the numerator.
The economic intuition for $dR/d\lambda \gtrless 0$ if and only if $h\Lambda \gtrless h_\lambda X\Theta$ goes as follows. Assume that $h\Lambda > h_\lambda X\Theta$ which is equivalent to $\Lambda/\Theta > h_\lambda X/h$. This inequality means that the relative increase in harvest cost is less than the negative impact of the cost shock on the harvest volume. Consider an upward harvest cost push. The steady-state resource stock rises, if the stock rising reduction of harvest quantity is larger than the stock depressing rise of harvest cost. But Corollary 11.1 states that this is just the case for our specification of harvest cost:

Corollary 11.1. *For the specified harvest cost function 11.2, a harvest cost push leads to an increase in the steady-state resource stock:* $dR/d\lambda > 0$.

Proof. In order to sign $dR/d\lambda$, we substitute for the derivatives of function A, and use $\Lambda = -\partial X/\partial\lambda_{|XX} \cdot \Theta$. This yields

$$\frac{dR}{d\lambda} = \frac{(h_\lambda h_R - hh_{R\lambda}X)\left[A - \alpha_1\sigma(1+g_R - h_R X)R\right]}{\Delta},$$

with $\left[A - \alpha_1\sigma(1+g_R - h_R X)R\right]/\Delta > 0$. For the chosen functional form for the harvest cost function 11.2, we clearly know that $(h_\lambda h_R - hh_{R\lambda}X) > 0$ and thus $dR/d\lambda > 0$. □

Thus, according to Corollary 11.1 and illustrated by Fig. 11.1, a harvest cost push increases the steady state resource stock and hence the new steady-state level R' is larger than the old one R.[2]

To provide some geometrical intuition for the sign of $dX/d\lambda$, assume that the *RR*-locus would not change due to the cost shock (i.e., $\partial R/\partial\lambda_{|RR} = 0$). Then, $dX/d\lambda$ is clearly positive. On the other hand, if the *XX*-locus would not respond to an increase in λ ($\partial X/\partial\lambda_{|XX} = 0$), $dX/d\lambda$ is negative because the stock is severely depleted if the harvest decision does not take account of the cost push. In the intermediate case, these two effects work in opposite directions, but for reasonable parameter values, the latter effect will outweigh the former such that $dX/d\lambda < 0$, as depicted in Fig. 11.1 where for the new steady-state harvest level $X' < X$ holds.

[2] The graphical illustrations require a numerical specification of the model. The parameters are chosen, in accordance with Props. 10.1 and 10.5, as $\alpha_1 = 0.05, \alpha_2 = 0.65, \beta = 0.9, \lambda = 30, \rho = 1, \pi = 0.86, R_{max} = 500$.

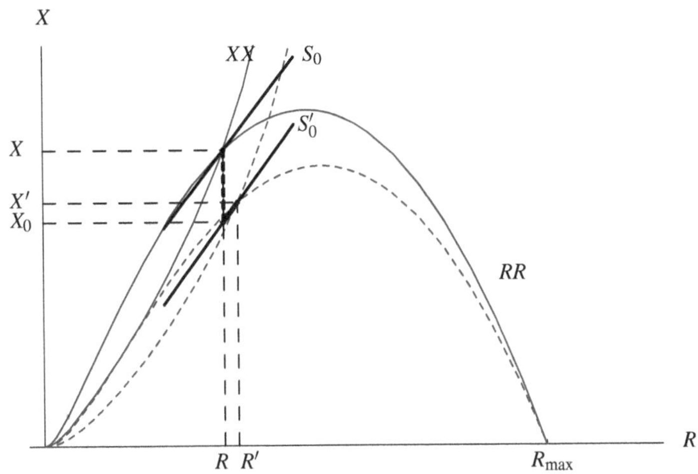

Fig. 11.1 Steady state and transitional effects of an increase in physical harvest costs

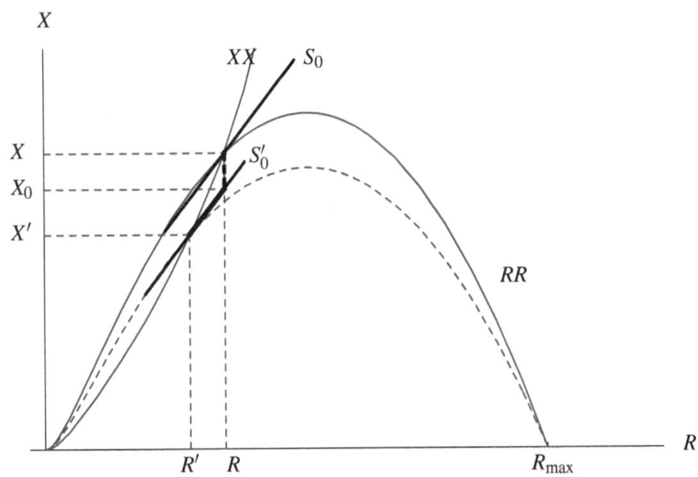

Fig. 11.2 Steady state and transitional effects of a decrease in the natural growth rate

11.2.2 Biological Shock: Lower Natural Growth Rate

Let us now turn to a biological shock, namely a shock to the resource growth rate. For that purpose, we have to take into account that the natural growth function depends on R and π and thus its partial derivatives will be denoted as: $\partial g(R,\pi)/\partial R \equiv g_R$ (instead of g') and $\partial g(R,\pi)/\partial \pi \equiv g_\pi$ etc. Prop. 11.2 gives the steady-state effects of a shock in the natural growth rate.

Proposition 11.2 (Steady state effects of biological shock). *The impact of an incremental change in the natural growth rate π on the steady-state value of R is given by:*

$$\frac{\mathrm{d}R}{\mathrm{d}\pi} = \frac{\mathrm{h}\Psi + \mathrm{g}_{\pi}\Theta}{\Delta} = \frac{\Theta}{\Delta}\left[\mathrm{g}_{\pi} - \mathrm{h}\left.\frac{\partial X}{\partial \pi}\right|_{XX}\right], \tag{11.6}$$

with $\partial X/\partial \pi_{|XX} \equiv -\Psi/\Theta$, *and* $\Psi \equiv \left[\mathrm{A}\mathrm{g}_{R\pi} + (1 + \mathrm{g}_R - \mathrm{h}_R X)\frac{\partial A}{\partial \pi}\right]$.
The impact on equilibrium resource harvest is determined by:

$$\frac{\mathrm{d}X}{\mathrm{d}\pi} = \frac{\mathrm{g}_{\pi}\Omega + (\mathrm{g}_R - \mathrm{h}_R X)\Psi}{\Delta} = \frac{\mathrm{g}_{\pi}\Omega - (\mathrm{g}_R - \mathrm{h}_R X)\left.\frac{\partial X}{\partial \pi}\right|_{XX}\Theta}{\Delta}. \tag{11.7}$$

If $\partial X/\partial \pi_{|XX} < 0$, *then* $\mathrm{d}R/\mathrm{d}\pi > 0$ *and* $\mathrm{d}X/\mathrm{d}\pi > 0$.

Proof. As before, $\Delta > 0$ and $\Theta > 0$. The signs of $\mathrm{d}R/\mathrm{d}\pi$ and $\mathrm{d}X/\mathrm{d}\pi$ can unambiguously be determined except with respect to Ψ. However, for a decrease in π, if the XX-locus gets steeper (turns counterclockwise) such that $\partial X/\partial \pi_{|XX} < 0$, then $\mathrm{d}R/\mathrm{d}\pi > 0$ and $\mathrm{d}X/\mathrm{d}\pi > 0$. □

The case $\partial X/\partial \pi_{|XX} < 0$ is depicted in Fig. 11.2 to illustrate the steady-state effects of a reduction in the natural growth rate of the resource stock ($\pi \downarrow$). The new *RR*-locus is below the previous *RR*-locus. The XX-locus turns counterclockwise (for the chosen parameter values, this turn can hardly be observed, however). The cumulative impact of a negative exogenous shock to the natural growth rate on the equilibrium levels of R and X is a decrease in both steady state values.

11.3 Transitional Dynamics towards the New Steady State

To be able to study the transitional dynamics following a parameter shock in the resource technology, the type and the timing of the shock must be specified. To keep the analysis as simple as possible we assume that the shocks in the resource technology parameters are permanent, unannounced and occur at the beginning of the transition period. While we focus in this section on the algebraic analysis of the two types of shocks, we provide in the Appendix a tool for numeric analysis. This computable general equilibrium (CGE) version of our model, implemented in the GAMS software package, can be utilized to perform additional parameter shocks and discuss the consequences for the three fundamental variables (R, X, K), as well as for all other economic quantity and price variables.

11.3.1 The Harvest Cost Push and the Overshooting of the Initial Harvest Quantity

To analyze the instantaneous jump of the resource harvest, we use the equilibrium manifold of the resource harvest as given in Prop. 10.2 to investigate the nature and the magnitude of the jump in resource harvest. In Fig. 11.1, the initial and the new equilibrium manifolds are denoted by S and S' respectively. Starting from the initial steady state, the resource harvest falls sharply in the first period onto the new equilibrium manifold S'_0 and then increases monotonically towards its new, steady-state level S' which is lower than before the shock. Since the resource stock is fixed in the shock period $t = 0$, it starts to grow in the post shock period $t = 1$ until it reaches its new higher steady-state level. Prop. 11.3 states analytically this result:

Proposition 11.3. *If $dX/d\lambda < 0$, then the resource harvest in the shock period $t = 0$ overshoots its new steady-state value:* $|\mathrm{d}X_0/\mathrm{d}\lambda| > |\mathrm{d}X/\mathrm{d}\lambda|$.

Proof. Using the equilibrium dynamics of the resource harvest (10.28a),

$$X_t = X + \frac{\overbrace{[J_{22}(R,X) - \psi_2]}^{>0}}{\underbrace{-J_{21}(R,X)}_{>0}}(R_t - R),$$

the initial adjustment in the harvest volume is given by

$$\frac{\mathrm{d}X_0}{\mathrm{d}\lambda} = \frac{\mathrm{d}X}{\mathrm{d}\lambda} - \frac{[J_{22}(R,X) - \psi_2]}{-J_{21}(R,X)}\frac{\mathrm{d}R}{\mathrm{d}\lambda} < \frac{\mathrm{d}X}{\mathrm{d}\lambda}.$$

Whether the change of X_0, the harvest volume in the shock period, is larger or smaller than the change of the harvest quantity in the new steady state (X) depends on the sign of the coefficient in front of $(R_t - R)$. Since this coefficient as well as $\mathrm{d}R/\mathrm{d}\lambda$ are certainly positive, the resource harvest in the shock period declines more than in the new steady state, thus $|\mathrm{d}X_0/\mathrm{d}\lambda| > |\mathrm{d}X/\mathrm{d}\lambda|$. □

To explain the economic rationale behind the sharp initial downward jump of the resource harvest we assume on the contrary that the resource harvest in the shock period does not change ($X_0 = \bar{X}$) and are led to a contradiction. Furthermore, we know that the stock variables do not change in the initial period: $R_0 = \bar{R}$, $K_0 = \bar{K}$. In view of the production function (left hand side of 10.13), the product supply remains unchanged. Then, the profit maximization conditions (10.12) imply that neither the price of the resource harvest nor the wage rate nor the real interest rate of the shock period change.

From the no–arbitrage conditions (10.8) and (10.9) it is easy to deduce that the price of the resource stock in the shock period (p_0) depends on the price of the resource harvest, on the own rate of return to the resource stock, and inversely on marginal harvest costs ($\mathrm{h}(\bar{R})$):

$$p_0 = \frac{\bar{q}(1 + \mathrm{g}'(\bar{R}) - \mathrm{h}'(\bar{R})\bar{X})}{\mathrm{h}(\bar{R})}.$$

Calculation of $\mathrm{d}p_t/\mathrm{d}\lambda$ shows that the increase in the marginal harvest costs is larger than the rise of the own rate of return, causing a fall in the initial stock price of the resource. Hence, the revenues of the old household from selling the natural resource stock decrease while the revenues from the ownership of real capital do not change. As a consequence, the consumption of the old household decreases in the shock period and with slight changes, the same applies to the consumption of the young household. For the market clearing condition (10.13) to hold, the fall in consumption of both young and old households requires that next–period capital stock increases. But a larger capital stock induces a lower interest factor. To ensure equal rates of return between real capital and the resource stock, the rate of return on the resource stock has to decrease too, which implies that its price has to decrease. But a fall in its price requires an increase in the resource stock R_1 which can be achieved only by a fall in the harvest level. Hence, the assumption of an unchanged resource harvest in the shock period is untenable and, instead of that, X_0 has to fall. Since there is a negative relationship between the resource harvest and its price (as well as the price of the resource stock), resource prices have to increase.

11.3.2 A Shock to the Resource Growth Rate

As for the harvest cost push, the coefficient in front of $(R_t - R)$ in (10.28a) is positive. However, due to the fact that the stationary resource declines in response to a negative shock in the biological parameters, the resource harvest volume of the shock period drops less than its new stationary–state value, a result that is contrary to the physical harvest cost shock where overshooting occurred. This is graphically illustrated in Fig. 11.2 where X_0 is smaller than its initial steady-state level X but larger than its new level X' and analytically shown in Prop. 11.4:

Proposition 11.4 (Transitional dynamics of biological shocks). *If* $\mathrm{d}X/\mathrm{d}\pi > 0$, *then the resource harvest in the shock period* $t = 0$ *does not overshoot its new steady-state value:* $\mathrm{d}X_0/\mathrm{d}\pi < \mathrm{d}X/\mathrm{d}\pi$.

Proof. Utilizing again the equilibrium dynamics of the resource harvest (10.28a), the resource harvest in the shock period 0 increases less than in the new steady state:

$$\frac{\mathrm{d}X_0}{\mathrm{d}\pi} = \frac{\mathrm{d}X}{\mathrm{d}\pi} - \frac{[J_{22}(X,R) - \psi_2]}{-J_{21}(X,R)}\frac{\mathrm{d}R}{\mathrm{d}\lambda} < \frac{\mathrm{d}X}{\mathrm{d}\pi}.$$

□

The economic intuition of the result is as follows. After a negative shock to the resource growth rate, the harvest volume is reduced in the shock period while the initial resource stock and the capital stock remain unchanged. Moreover, in the shock period the price of the harvest and of the resource stock increase, while the rate of return on the capital stock falls slightly. As a consequence, for the young household the income from working and resource harvest falls in the shock period, leading to a fall in consumption. On the other hand, for the old household income and consumption increase if the rate of return from the resource stock increases by more than the fall in the rate of return from the capital stock. In the following period, however, the resource stock falls on account of higher harvest costs and this process continues until the new, lower steady-state level is reached.

11.4 A Comparison of a Harvest Cost Push and a Biological Shock

Geometrically, the differences between a harvest cost shock and a biological shock can be explained as follows. First, a positive harvest cost shock ($\lambda \uparrow$) leads to an increase in the steady-state resource stock (see the downward shifts of both loci in Fig. 11.1). Thus, a harvest cost shock leads to a resource recovery due to the substitution of labor and capital for harvest input. On the other hand, a negative biological shock implies a downward shift of the *RR*-locus but an upward shift of the *XX*-locus (see Fig. 11.2). Accordingly, the stationary–state resource stock and the harvest volume decline. On interpreting the shift of the *XX*-locus, we find that households substitute man-made capital for the resource stock since the resource stock becomes less profitable.

Second, the transition paths, as indicated by the bold dotted lines in Figs. 11.1 and 11.2, are driven by the position of the initial and new steady states as well as the shift of the stable arm from S_0 to S_0'. For the harvest cost shock, the shock leads to a downward shift of the positively sloped stable arm. Since the new steady state is to the right of the initial one, harvest falls to the new stable arm S_0' and moves upward along the arm towards the new steady state. On the other hand, a biological shock leads to a new steady state resource stock to the left of the initial stock. Thus, harvest drops to the new stable arm and moves downward along the arm to the new lower steady state.

The economic intuition why overshooting occurs for a harvest cost push but not for a shock in the natural growth rate is as follows. For the case of the cost shock, there is a direct impact of the cost push on the harvest level, leading to a sharp initial drop in the harvest volume and a corresponding rise in the harvest price. As a result, the price of the resource stock rises in the shock period, too. In the long run, however, the resource stock recovers to a new higher steady-state level and the new steady-state harvest volume is moderately lower than in the pre–shock equilibrium. Thus, the harvest volume initially overshoots its new steady-state level and so does its price. Moreover, in the new steady state man-made capital, total output and consumption of both young and old households are lower than in the pre–shock equilibrium.

On the other hand, a negative shock in the resource growth rate has a direct impact on the resource stock and only an indirect one on the harvest volume. Thus, in the post–shock period the harvest volume is only moderately reduced, its price increases moderately and the price of the resource stock increases moderately, too. As for the harvest cost push, consumption of the young household declines while consumption of the old household increases in the shock period. Thereafter, however, both the resource stock and the harvest volume degrade gradually towards their new lower steady state levels. As before, in the new steady state man-made capital, total output and consumption levels of young and old households are lower than in the initial steady state.

11.5 Conclusions

In this chapter we investigated the potentially different impacts of a harvest cost and of a biological shock, both in the long run as well as along the transition path towards the new stationary state. As regards the steady-state effects of a harvest cost push we found a positive impact on the resource stock whereas the resource harvest usually decreases. That the steady-state resource harvest declines while the resource stock increases is also the main reason why the resource harvest of the shock period overshoots its new stationary state value. As a consequence, the price of the resource harvest and the price of the resource stock overshoot, too.

A negative shock to the natural growth rate has a negative impact on the main dynamic variables in the new steady state. In view of the equilibrium manifold (the stable arm) the resource harvest does not overshoot its new, lower steady-state value. This result can be seen as a rationale for modest short run economic responses, observed in real world circumstances, to shocks in nature's technology in spite of their much larger long-run impacts.

Three directions for future research are easily identified. First, other functional forms for the resource harvest could be used, e.g. relaxing the assumption of harvest costs linear in the harvest volume. Another option were to replace the inverse impact of the resource stock such that harvest costs increase with the resource stock, a specification suitable e.g. for species–rich ecosystems like tropical forests. Second, the notion of physical harvest cost could be replaced by harvest costs in terms of labor (see Appendix A to Chap. 10 in the log-linear CD OLG model), and the impacts of shocks to the resource technology analyzed. Third, rather than focusing on log–linear intertemporal utility functions a more general function (like CES) should be considered. The equilibrium dynamics would then depend on the man-made capital in a non-trivial way, and a richer and hence empirically more interesting dynamical system is to be expected.

Appendix: GAMS Code to Analyze the Model

The GAMS code starts with a declaration of variables and parameters:

```
$TITLE reduced form OLG-CGE model with resources.

SETS
        t     time periods /1*100/;

SET     TFIRST(t), TLAST(t);

        TFIRST(t) = YES$(ORD(t) EQ 1);
        TLAST(t) = YES$(ORD(t) EQ CARD(t));

ALIAS   (t,TT) ;

SCALARS

    ss    savings rate
    bb    time discount factor                  /0.9/
    a1    prod. elasticity of resource harvest /0.05/
    a2    prod. elasticity of labor             /0.65/
    a3    prod. elasticity of capital           /0.3/
    rr    growth factor of the resource         /0.72/
    RM    growth retarding factor               /500/
    L     scaling factor in harvest cost        /30/
    gg    marginal propensity to consume
    u     level parameter in production         /100/;

PARAMETERS

    XZ    Resource harvest (initial value)
    RZ    Resource stock (initial value)
    KZ    Capital stock (initial value);

    ss=bb/(1+bb);
    gg=1-ss;

    XZ = 84.82952741;
    RZ = 1.794297E+2;
    KZ = 1.189157E+2;

VARIABLES
    X(t)    Resource harvest in period t
    R(t)    Resource stock in period t
    K(t)    Capital stock in period t
    G(t)    Regeneration function in t
    GR(t)   Regeneration function R derivative in t
    H(t)    Harvest cost function in t
    HX(t)   Harvest cost function X derivative in t
    HR(t)   Harvest cost function R derivative in t
    NH(t)   Net harvest function derivative in t
    A(t)    A function
    F(t)    F function
```

```
     Y(t)     Aggregate Output
     q(t)     Harvest price
     w(t)     Wage rate
     r(t)     Capital service price
     p(t)     resource stock price
     C1(t)    Young consumption
     C2(t)     Old consumption
     YS(t)     Product excess supply
     TRICK     Objective variable;

      POSITIVE VARIABLES X,R,K;
```

The next step is to define the model equations:

```
EQUATIONS

     EQX(t)       harvest dynamics
     EQR(t)       resource dynamics
     EQK(t)       capital dynamics
     EQG(t)       regeneration function
     EQGR(t)      regeneration function derivative
     EQH(t)       harvest cost function
     EQHX(t)      harvest cost function X derivative
     EQHR(t)      harvest cost function R derivative
     EQNH(t)      net harvest cost function derivative
     EQA(t)       A function
     EQF(t)       F Function
     EQY(t)       aggregate production function
     EQq(t)       harvest price equation
     EQw(t)       wage rate equation
     EQr(t)       capital service price equation
     EQp(t)       resource stock price equation
     EQC1(t)      young consumption equation
     EQC2(t)      old consumption equation
     EQMC(t)      product excess supply equation
     EQXL(t)      terminal condition X
     EQRL(t)      terminal condition R
     EQKL(t)      terminal condition K
     EQMCL(t)     terminal condition MC
     EQOBJ        Objective function (TRICK variable);
```

The three dynamic equations read as follows:

```
*First dynamic equation

EQR(t+1)..     R(t+1)=E=R(t)+ G(t)-H(t);

EQRL(TLAST).. R(TLAST)=E=R(TLAST)+ G(TLAST)-H(TLAST);

*Second dynamic equation

EQX(t+1).. X(t+1)=E=A(t)*(1+rr-2*(rr/RM)*(R(t)+G(t)-
           H(t)))/(a3*(1+L/(1+R(t)+G(t)-H(t)))-A(t)*
           L/(1+R(t)+G(t)-H(t))**2);
```

```
EQXL(TLAST).. X(TLAST)*(a3*(1+L/(1+R(TLAST)+G(TLAST)-
              H(TLAST)))-A(TLAST)*L/(1+R(TLAST)+
              G(TLAST)-H(TLAST))**2)=E=A(TLAST)*
              (1+rr-2*(rr/RM)*(R(TLAST)+G(TLAST)-
              H(TLAST)));

*Third dynamic equation

EQK(t+1).. K(t+1)*(1+GR(t+1)-HR(t+1))*HX(t)=E=
           u*a3*X(t+1)*HX(t+1)*X(t)**(a1-1)*K(t)**a3;

EQKL(TLAST).. K(TLAST)*(1+GR(TLAST)-HR(TLAST))*HX(TLAST)
             =E=u*a3*HX(TLAST)*X(TLAST)**a1*K(TLAST)**a3;
```

The remaining equations are auxiliary equations to determine harvest costs, prices, consumption quantities etc.:

```
*2. Other model variables:

EQG(t)..       G(t)  =E= rr*R(t)-(rr/RM)*R(t)**2;

EQGR(t)..      GR(t)  =E= rr-2*(rr/RM)*R(t);

EQH(t)..       H(t)  =E= X(t)+(L*X(t)/(1+R(t)));

EQHX(t)..      HX(t)  =E= 1+(L/(1+R(t)));

EQHR(t)..      HR(t)  =E= -L*X(t)/(1+R(t))**2;

EQNH(t)..      NH(t)  =E= (1+GR(t)-HR(t))*R(t);

EQF(t)..       F(t)  =E= G(t)/R(t)-GR(t);

EQA(t)..       A(t)=E=(a1+a2*ss)*X(t)*HX(t)-a1*gg*(F(t)+
               HR(t))*R(t)-a1*(1+GR(t)-HR(t))*R(t);

EQY(t)..       Y(t)=E=u*X(t)**a1*K(t)**a3;

EQq(t)..       q(t)=E=u*a1*X(t)**(a1-1)*K(t)**a3;

EQw(t)..       w(t)=E=a2*Y(t);

EQr(t)..       r(t)=E=u*a3*X(t)**a1*K(t)**(a3-1);

Eqp(t)..       p(t)*HX(t)=E=q(t)*(1+GR(t)-HR(t));

EQC1(t)..      C1(t)=E=gg*(w(t)+(q(t)*(1+R(t))/
               (1+R(t)+L))*(G(t)-GR(t)*R(t)+HR(t)*R(t)));

EQC2(t)..      C2(t)=E=r(t)*K(t)+(q(t)*(1+R(t))/
               (1+R(t)+L))*(1+GR(t)-HR(t))*R(t);

EQMC(t+1)..    YS(t)=E=Y(t)-C1(t)-C2(t)-K(t+1);
```

```
EQMCL(TLAST)..  YS(TLAST)=E=Y(TLAST)-C1(TLAST)-C2(TLAST)-
                K(TLAST);
```

The optimization for a discrete dynamic model is invoked by:

```
* Objective

EQOBJ..             TRICK =E= 1;

MODEL OLGCGE2 /ALL/;
```

To ensure that the numerical routine succeeds in finding a solution, initial and boundary conditions are defined:

```
* Initial values and boundaries

X.L(t) = XZ; R.L(t) = RZ; K.L(t) = KZ;

X.LO(t) =.000000001;
R.LO(t) =.000000001;
K.LO(t) =.0000000001;


* Initial stocks

R.FX(TFIRST) = RZ; K.FX(TFIRST) = KZ;

OPTION decimals=8;
```

The model solver is initiated by the following statement:

```
SOLVE OLGCGE2 MAXIMIZING TRICK USING NLP;
```

To read the solution for the three dynamic variables as well as for factor prices, consumption and output levels, as well as harvest cost, the following command is used:

```
DISPLAY
X.L,R.L,K.L,Y.L,q.L,p.L,w.L,z.L,C1.L,C2.L,YS.L,HX.L,
HR.L,NH.L;
```

References

Gandolfo G (1997) Economic Dynamics. Springer, Berlin, Heidelberg
Samuelson P (1947) Foundations of Economic Analysis. Harvard University Press, Harvard

Index

FSC
www.fsc.org
MIX
Papier | Fördert gute Waldnutzung
FSC® C083411

Kolibri 360 GmbH
Ferdinand-Jühlke-Straße 7
99095 Erfurt, Deutschland
produktsicherheit@kolibri360.de